Now with Added Algorithms

A Guide for Parents and Families in Australia's AI Future

SHELLY DUNCAN

Now with Added Algorithms
A Guide for Parents and Families in Australia's AI Future

Published by Wyld & Ink Creative Press

First published in Australia in 2025.

ISBN: 978-1-7640456-1-2

Cover image generated with ChatGPT AI; design by Shelly Duncan

This book is for informational purposes only. While every effort has been made to ensure accuracy at the time of publication, the field of artificial intelligence is rapidly evolving. Readers are encouraged to consult current sources for the most up-to-date information.

Dedication

To my husband — who was always on call to keep our five-year-old daughter entertained when she had very helpful plans for my writing time.

To our daughter — for being the heartbeat behind every chapter. She is the reason I wanted to learn more, understand better, and stay curious.

And to those who've gone before — the educators, researchers, and advocates who've long worked to make technology more inclusive, thoughtful, and human. This work stands on your shoulders.

Table of Contents

Introduction:
Why This Matters Now

"What kind of future are we preparing our kids for — and how can we help them thrive in a world shaped by AI?"

Not long ago, teenagers across Australia would get their first taste of independence behind the counter of a fast-food chain — juggling orders, learning to smile through pressure, earning that first payslip. But now?

At a fast-food outlet in Sydney's inner west, the voice taking your order isn't a young worker anymore — it's artificial intelligence.

Imagine your teen, excited to get that first pay, heading down to the local takeaway. They've heard stories from older siblings or even parents about the buzz of a Friday night shift — juggling orders, remembering who wanted extra pickles, trying to smile through the chaos. But now, that very first step into working life is being handled by something else: a polite, tireless, ultra-efficient AI.

You may have seen clips online — AI systems taking drive-through orders without missing a beat. No hesitation. No errors. No "Just give me a second." Just smooth, unshakable efficiency.

At first glance, it might seem like a small change. But what's lost isn't just a bit of pocket money — it's a rite of passage. These were the roles where young people learned how to work with others, solve problems on the fly, manage time, and build confidence. And as more of these entry-level jobs disappear, the pathway into adulthood gets a little blurrier.

Moments like this spark something in the national conversation — especially among parents. What happens when the kinds of jobs we once relied on to teach our kids responsibility and resilience start to disappear?

And it's not just retail. AI is being trialled in classrooms, replacing customer service teams, and handling everyday decisions in banking and service agencies. It's supporting healthcare workers, screening resumes, helping create media, and even assisting in government services. Often, we don't even realise it's happening — the technology blends in quietly, behind the scenes.

We encounter it daily, whether we know it or not — in the way Google Maps reroutes us, Netflix recommends our next show, or Gmail suggests what we might want to say next. AI isn't coming — it's already here. In small, persistent ways, it's becoming part of how we work, learn, shop, scroll, and communicate.

And while many of us are still trying to catch up — to figure out what AI even is — our kids are already growing up with it shaping their learning, entertainment, and the very nature of the jobs they'll be aiming for. (Put simply: AI is software that can perform tasks that usually need human thinking — like recognising speech, writing text, or spotting patterns. It doesn't "understand" like a person does. It matches patterns, not meaning — and that's an important distinction we'll return to.)

And yet, for all this exposure, trust in AI remains low — especially here in Australia.

Recent national studies show Australians have one of the lowest levels of trust in AI globally. In a 2025 survey of over 1,000 business leaders, 83% believed AI would impact their operations in the next year — but only 39% felt confident implementing it. Another 2024 report from the federal government found that while digital tools are central to public services, trust and inclusion remain major gaps. These aren't just technical issues. They're human ones.

But if you've ever felt a bit wary, you're not alone — and you're not wrong to pause.

That low trust doesn't mean Australians are anti-tech. It means we tend to take a breath before diving in. We ask questions. We want to understand what's behind the curtain.

Take a common example. Imagine your child's school introduces a new tool that uses AI to give feedback on their writing. It sounds helpful — faster feedback, more personalised support. But as a parent, it's natural to pause.

You might find yourself wondering: What's actually happening with my child's work? Is the feedback accurate? Where is the data going?

That's not being difficult — that's being thoughtful. And that kind of instinct is something we should encourage. It's not about rejecting the tech. It's about staying curious, asking questions, and making sure that trust is something earned — not assumed.

In a world where AI can sound confident — even when it's wrong — that instinct is a strength. It can help us, and our kids, think critically. To ask: "Who built this?", "What's it trying to do?", "Is this helping or just automating for the sake of it?"

Trust shouldn't be automatic. It should come from understanding — from seeing that a tool is genuinely useful, safe, and aligned with how we want our families to live and learn. That kind of discernment isn't just important in the age of AI — it's essential.

This book isn't here to scare you. It's here to help. To offer calm, clear guidance in a time of fast change. You don't need to be a tech expert to help your child navigate the age of AI — but you do need to understand what's happening now, and what's coming next.

I see this shift from two angles.

I've spent my career working in the IT sector — mostly in marketing, partnerships, and strategy roles. That means I'm often at the interface between technology and people, helping organisations make sense of what's changing, how to communicate clearly, and how to adapt when everything's moving fast. I'm not building these tools myself, but I see how they're being used — and how quickly they're becoming part of everyday life.

I use AI in my work every day — not in theory, but in the flow of real tasks: research, writing, planning, problem-solving. It's useful, yes — but I also see where it falls short. How easily it can overwhelm or oversimplify. How tempting it is to hand over decisions we should still be making for ourselves.

At the same time, I'm especially interested in the good technology can do — the ways it can support inclusion, wellbeing, and creativity. But only if we stay thoughtful in how we use it. And that's really what this book is about: not the tech itself, but how we stay human at the core.

That professional lens is one part of it. The other is much closer to home — I'm a parent with a young daughter. She's bright, curious, and already asking big questions about the world. Like many parents, I find myself thinking about the future she's growing into — not just in terms of careers or education, but the values and skills that will help her live a meaningful life.

Hers is a world where machines won't simply support her learning — they'll influence how she sees herself, what she believes is normal, and how she measures success. These systems are shaping her generation's sense of identity, self-worth, and possibility.

That's both exciting and confronting.

What kind of future are we preparing our kids for — and how can we help them thrive in a world shaped by AI?

That question sits at the heart of this book. It's what drives the conversations we need to be having — not just about technology, but about values, decisions, and the kind of adults we're hoping to raise.

I want her to grow up confident with technology — not unquestioning of it.

I want her to feel capable using digital tools, but also to pause and think critically about how they work, who built them, and what they're designed to influence. Being tech-savvy is important — but so is knowing when to step back, ask questions, or choose a different path. Confidence shouldn't come at the cost of curiosity or discernment.

And just as importantly, I want her to know that qualities like imagination, empathy, and integrity matter every bit as much — if not more — than the ability to master any new tool. These are our human superpowers. The ones machines can't replicate. And the ones our kids will need to draw on, again and again, to shape the kind of future they want to live in.

That balance — between capability and discernment, between what's possible and what's purposeful — is becoming harder to hold. The pace of change makes it tempting to either hand over control or step back entirely. But neither option serves our kids. And that's why this conversation is so important.

You don't need to be an expert in artificial intelligence to guide your child through it.

You just need to stay present. Curious. Willing to ask questions, to think critically, and to hold onto the human qualities that machines can't replicate — the ones your child needs most.

This isn't about keeping up with every new technology.

It's about staying grounded in the kind of person you're raising — and the kind of future you want them to help shape.

The time to engage isn't later. It's now.

A note on change

The world of AI is evolving fast — so fast, in fact, that by the time you're reading this, some of the tools or examples in this book may already have changed. New platforms launch, policies shift, and technology keeps moving.

But this book isn't just about keeping up with the latest app or algorithm.

It's about something more lasting: the human side of parenting in a digital world. Curiosity. Connection. Asking good questions. Helping our kids make sense of the tools around them — not just use them.

Those things don't go out of date.

Chapter One: The Age of AI Is Already Here

"We know AI is changing things — but what is it, really? And how is it already reshaping the world our kids are growing up in?"

You might not think of yourself as someone who uses artificial intelligence. But chances are, it's already woven into your daily routine.

It's there when Netflix recommends your next show, when Google Maps suggests a faster route, when your email drafts a reply before you've even finished reading. It's behind voice assistants, autocorrect, personalised ads, chatbots, online banking, and even the spam filter keeping your inbox sane.

AI is no longer a "what if." It's a "right now."

And while the headlines often focus on extremes — from robots taking over jobs to AI writing novels or passing medical exams — the truth is often quieter. More ordinary. And more deeply embedded in how we live, learn, work, and parent.

So before we go further, let's answer the question many people are still asking.

What is AI, really?

You've probably heard the term "AI" more in the last year than in the last decade — on the news, in your workplace, maybe even in your child's classroom. But what is it really?

For most of us, it still feels vague. Not because we're behind, but because the people building these systems often explain them in technical language that leaves the rest of us guessing.

In simple terms, AI (artificial intelligence) refers to software that can perform tasks that usually require human thinking — such as recognising speech, translating language, making predictions, or creating content.

But that doesn't mean it thinks the way we do. Far from it.

What it has is speed, memory, and an enormous ability to process patterns.

A common type of AI we hear about today is a large language model — like ChatGPT. These are trained on vast amounts of text from books, articles, websites — and they learn to predict what words should come next in a sentence. That's why AI can now write essays, emails, or poems that sound intelligent. But under the hood, it's still pattern-matching — not thinking.

A helpful way to picture AI is as a super-fast library assistant. One who's read everything ever published online, can retrieve and combine that information in an instant — but has no lived experience, no emotion, and no sense of context. It doesn't know what love feels like. It doesn't understand grief. It doesn't know what it means to grow up, to be excluded, or to lose a job. It doesn't have beliefs. It doesn't form opinions. It doesn't care about being right — only about being convincing.

That's not a flaw. That's how it's designed.

Useful? Yes. Impressive? Often. But human? Not even close.

Have you met AI? (spoiler: you already have)

If you've used:

- ✓ Google Maps to dodge traffic,
- ✓ Netflix to find your next show,
- ✓ Online shopping sites that somehow know what you want before you do,
- ✓ A chatbot while trying to sort out your internet bill, or
- ✓ Gmail's eerily accurate "suggested replies"

...you've already met AI. And almost without doubt, so has your child.

AI isn't some futuristic gadget waiting in the wings. It's already here — quietly shaping how we live, learn, and communicate. It's the invisible helper behind the scenes, streamlining our routines, guiding our choices, and occasionally finishing our sentences.

It's not just knocking at the door anymore. It's wandered in, kicked off its shoes, and started helping with the homework.

And that's what makes this moment both exciting and urgent. Because if AI is already part of family life — whether we've invited it or not — then understanding how it works, where it helps, and where we need to stay alert becomes part of modern parenting.

How AI tools learn — and why that matters

Most of the tools we now call AI are powered by something called machine learning. Instead of following fixed rules, these systems are trained on huge volumes of data — text, images, code, audio, video — to detect patterns and generate responses.

In other words, they don't "know" things. They predict what's most likely to come next, based on what they've seen before.

That prediction engine is incredibly powerful. But it also means AI picks up not just facts, but also opinions, assumptions, and biases buried in its training data. Because most of this data comes from the internet — and the internet reflects society — AI ends up absorbing all the beauty, creativity, messiness, and unfairness we've published over time.

That has real-world consequences.

Imagine asking an AI to describe a "typical parent." The answer it gives depends entirely on what examples it's been exposed to. If the training data includes mostly American blogs, English-language forums, or corporate articles, the AI may give a skewed picture — one that misses cultural context, diverse family structures, or lived experience outside the mainstream.

And because AI often sounds confident — even when it's wrong — it can be easy to trust what it says at face value.

This is especially important when our kids start using these tools.

They might get different results from a classmate using the same app.

They might see suggestions that feel "off" — but not know why.

They might assume the tool knows best — when really, it's just echoing what it's been fed.

That's why it helps to pause and ask:

- Where did this information come from?
- Who gets included — and who's missing?
- What might this tool not understand?

Because teaching our kids to be thoughtful users of AI starts with understanding that these tools are not neutral. They're powerful mirrors of the data they're trained on — and like any mirror, they can reflect some things clearly, and distort others.

Pop culture vs. reality

Wait – Isn't AI Supposed to Be a Robot?

When we think of AI, many of us picture the movies: robots taking over, glowing red eyes, cold machine voices with a mind of their own. From The Terminator to Ex Machina, pop culture has trained us to expect something dramatic – and a little terrifying.

But real AI? It's far more subtle.

- ✓ It's the chatbot helping with homework
- ✓ The algorithm recommending your next YouTube video
- ✓ The voice assistant setting a timer for dinner

There's no emotion, no secret plan – just a system trained to spot patterns and make predictions.

The real challenge isn't science fiction.

It's how these tools shape what our kids learn, see, and believe – quietly, every day.

Letting go of the robot myth helps us stay grounded in what matters: raising kids who can think for themselves, even in a world full of smart machines.

It's not the future – it's now

For years, artificial intelligence was something we associated with science fiction – the kind of tech that lived in the far-off future, wrapped in robot takeovers and dystopian plotlines. Most of us didn't think about it in practical terms. It was something for researchers, for Hollywood scripts, for "someday."

But then, something changed.

In late 2022, an AI tool called ChatGPT was released to the public. At first, it was mostly explored by tech enthusiasts and early adopters. But in the years since, its reach has expanded dramatically — into classrooms, workplaces, and homes across Australia and beyond. People now use AI to write emails, generate lesson plans, assist with homework, brainstorm business ideas, and even answer parenting questions. It hasn't just entered the mainstream — it's steadily reshaping how we think, work, and relate to information.

The big shift wasn't just what AI could do — it was who could use it, and how easily. For years, AI sat in the background of big tech — powering maps, playlists, and shopping recommendations without much notice. But when tools like ChatGPT became available to everyone, that changed. Suddenly, we weren't just surrounded by AI — we could talk to it, experiment with it, even ask it parenting questions. It moved from something only tech companies controlled to something we could all interact with, right from our phones or laptops. That kind of access changed everything — especially for families.

And for our kids, that access feels completely normal. They're growing up in a world where asking a chatbot for help is as ordinary as using a calculator or watching a YouTube tutorial. But just because they can use it easily doesn't mean they understand how it works — or what it's shaping underneath. The more embedded it becomes, the more it matters that we understand where and how it's being used — especially when it touches our kids' lives.

Australia's growing AI industry

Australia's AI ecosystem is expanding rapidly — and gaining international attention:

- ✓ 1,533 AI companies are now operating in Australia, up from just 544 two years ago. That's a sharp rise, driven by genuine growth and better tracking. *Source: industry.gov.au*
- ✓ *Over $2 billion in venture capital was invested in Australian AI startups in 2023 alone.* Source: minister.industry.gov.au
- ✓ *AI-related patents have nearly quadrupled — from 170 in 2015 to 629 in 2024 — reflecting strong research momentum.* Source: industry.gov.au

But momentum isn't the full picture. While the sector is growing fast, Australia still ranks low in areas like public trust, digital skills confidence, and national AI strategy. In other words: we're building the tools — but not everyone feels ready to use them.

Helping kids understand and engage with AI isn't just about future jobs — it's about giving them the tools to thrive in a world already shaped by these systems.

But momentum isn't the full picture. While the sector is growing fast, Australia still ranks low in areas like public trust, digital skills confidence, and national AI strategy. In other words: we're building the tools — but not everyone feels ready to use them.

That's why parents matter in this story. Helping kids understand and engage with AI isn't just about future jobs — it's about giving them the tools to thrive in a world already shaped by these systems.

Right here in Australia, AI is already changing how we learn, care, work, and connect — often in ways that directly affect families. Some of these shifts are obvious, like chatbots in schools or smart assistants in our homes. Others are buried deep in the systems we interact with every day. But together, they're reshaping the world our children are growing up in.

Here's what that looks like today:

AI in Australian classrooms

In classrooms across Australia, AI is starting to lend a quiet hand — not to take over, but to give teachers back a bit of breathing room. Some schools are testing tools that help with planning, marking, or giving students quick feedback — the behind-the-scenes work that often eats up time.

Education Perfect ran a 10-week study with nearly 20,000 students and saw a 47% jump in the quality of student responses. In South Australia, a chatbot called EdChat is answering routine questions so teachers can spend more time with their students. And in Queensland, Brisbane Catholic Education is rolling out AI support across more than 140 schools to save teachers hours each week.

We'll explore more about what this means for students, teachers, and families in Chapter 3 — but even now, it's clear that AI is starting to shape the classroom in quiet but significant ways.

AI in healthcare

Australian hospitals are adopting AI to enhance diagnostic processes and patient care. At the Gold Coast University Hospital, AI is starting to help speed up how scans are reviewed — things like X-rays, MRIs, and CTs. It's not replacing the experts, but it is helping them get through the pile a little faster.

And for families, that matters. Because when you're waiting on scan results — especially for your child — time stretches. You check your phone. You check again. You wonder if no news is good news.

AI doesn't take that worry away, but it can help move things along. It can flag urgent cases faster or help doctors spot what needs a closer look. That means less waiting, quicker answers, and hopefully, earlier treatment when it's needed.

It might not grab headlines — but it's one of those quiet ways AI is already helping in the background, making hard moments just a little easier.

AI in banking

Managing family finances already takes enough headspace — bills to pay, transfers to make, kids needing pocket money (again). The last thing any parent wants is to fall for a scam or send money to the wrong place.

That's why banks like the Commonwealth Bank are building AI into their everyday tools. Features like NameCheck give you a heads-up if something doesn't match before you hit "send." CallerCheck can help you figure out whether a call is really from the bank — or someone pretending to be.

You might not even notice these tools are powered by AI, but they're making a difference. CBA has reported a 50% drop in scam losses and a 30% fall in fraud reports from customers. That's huge — and for families, it means one less thing to stress about when moving money or managing accounts. It's not the kind of AI that writes essays or chats back — but it is helping keep your finances a little safer.

AI in recruitment

Across Australia, approximately 30% of businesses are now using AI in their hiring process — not just big tech firms, but everyday companies looking to save time and streamline recruitment (University of Melbourne, 2025). These tools are helping with things like scanning resumes, matching applicants to job descriptions, and even scheduling interviews.

In practice, here's what that can look like:

- Resume Screening: AI scans for keywords and qualifications that match what the employer's looking for.
- Candidate Matching: It predicts who might be a good fit based on previous experience or skill sets.
- Interview Scheduling: Some systems handle the back-and-forth of finding a time that works for everyone.

It's fast, efficient, and can take some pressure off hiring managers — especially when they're dealing with hundreds of applications.

But here's the catch: these systems aren't perfect. Studies show they can sometimes miss great candidates, especially if someone's experience doesn't follow a typical path. That could include young people with employment gaps, different learning journeys, or those who speak English as a second language.

It's not about turning back the clock. It's about making sure there's still a human in the loop — someone who can see potential that an algorithm might overlook.

Later in the book, we'll look more closely at where bias can creep into AI systems — and what parents and educators can do to help young people prepare for a job market that's still learning how to be fair.

AI in government services

You've probably heard about AI being used in Centrelink to flag potential fraud or manage debts. But that's just one example. Across Australia, governments are trialling AI in everything from waste collection to traffic lights.

In Brisbane, the Smarter Suburban Corridors project will test AI that adapts traffic light timings in real-time to ease congestion. In South

Australia, smart road cameras are adjusting signals based on traffic flow — and in Sydney, the long-running Sydney Coordinated Adaptive Traffic System (SCATS) quietly keeps cars moving by sensing vehicle patterns and responding on the fly.

It's not just the roads that are getting smarter. In NSW, local councils are rolling out AI-equipped garbage trucks and buses that can detect potholes as they do their rounds, making it easier to spot and fix hazards quickly. Smart bin technology is gaining traction in many of our capital cities to help optimise recycling, waste sorting and collection.

Even local planning offices are turning to AI. In NSW, several councils are using AI to streamline development applications, reducing delays and making the process smoother for residents and small businesses alike. Document scanning, categorising reports, even helping sort community requests through tools like the Snap Send Solve app — AI is making local services faster and more responsive.

AI in social media

Chances are, you already know that social media feeds aren't just a random scroll — they're powered by algorithms. What shows up on your screen isn't just what's "new" or "popular," but what AI thinks you'll engage with most. That includes posts from friends, trending videos, tailored ads — even those oddly accurate product suggestions that pop up minutes after you've mentioned something out loud.

But beyond curating our feeds, AI is now being used to create content, too. Meta — the company behind Facebook and Instagram — is rolling out features like "Imagine," which lets users generate images from simple text prompts. Think: "a birthday cake shaped like a koala" or "a futuristic city in the outback." These images can be shared directly in feeds, stories, or even used as profile pictures.

Meta is also experimenting with AI-generated users that can post, respond, and even hold conversations — all powered by synthetic personalities. And as generative content becomes more common, they've started labelling AI-created images to help people spot what's real and what's not.

It's not all art and fun, though. Ahead of elections and major news events, AI is increasingly part of the effort to detect deepfakes and misinformation before they spread too widely.

So even if you're not actively using AI tools yourself, chances are you're seeing — and being influenced by — AI every time you scroll.

From breakfast to bedtime: how AI shows up at home

AI isn't just something happening "out there" in offices, schools, or tech labs. It's already made its way into the heart of family life — often without us even noticing.

At home, voice assistants cheerfully answer questions about dinosaurs or the solar system while toast is popping and school bags are being packed. Educational apps offer just-right praise or extra practice when your child gets stuck. Video platforms line up the next clip — and the one after that — keeping little eyes glued to the screen for longer than anyone intended. Even those oddly accurate ads we scroll past? They're often shaped by something someone in the house said out loud.

None of this is random. It's AI, quietly working in the background — learning, predicting, nudging — without asking for permission or explaining how it works.

It's convenient. Sometimes even helpful. But it's also fast-moving and mostly invisible. And the people designing these systems? They don't know your child. They don't live in your house. And they're not there to explain what's going on behind the screen.

That's not a reason to panic. But it is a reason to stay curious and pay attention.

Because while many of us are still figuring out what AI is and how it works, our kids are already living with it. They don't see it as new or futuristic. For them, it's just... normal. Part of growing up.

And that's what makes this moment different.

If we want to raise thoughtful, capable kids in this environment, we can't just ask what AI can do. We need to ask how it's shaping the way our children think, learn, and make sense of the world.

Because while AI might help with homework or keeping them entertained, it's still our job to guide them in the things machines can't teach — like judgement, creativity, and knowing who they are.

> **How many Australians use AI?**
>
> Almost 1 in 2 adults in Australia reported using generative AI tools like ChatGPT in the past year — up from 38% in 2023.
> *Source: Google & IPSOS, 2025*

Why this matters for parents

While AI might help with homework or keep our kids entertained, it's still our job to guide them in the things machines can't teach — like judgment, creativity, and knowing who they are.

You don't need to become a technology expert to guide your child through this new world — but you do need to stay engaged.

And that doesn't require mastering every new tech trend. It just means staying curious. Paying attention to how these tools show up in your child's life — and what they might be reinforcing.

Because AI isn't just shaping what kids do. It's shaping how they think, how they compare themselves to others, and what they believe matters.

It's part of their environment now — like the schoolyard or the dinner table — only much of it happens behind the scenes.

That's why your role is so important. You're the one who can help connect the dots. To keep the focus on the human stuff — values, questions, boundaries, relationships. That's what really helps kids find their way, no matter how fast the tech moves.

> **Did you know?**
>
> Only 36% of Australian parents say they feel very prepared to talk to their kids about online risks — even as AI tools become part of daily life. *Source: Microsoft Global Online Safety Survey, 2025*

It's already shaping the way they think and learn

AI is already nudging how our kids learn, think, and even see themselves.

How they learn

AI is already baked into many of the tools kids use for school and homework — even if it's not always obvious. A maths app might adjust its questions on the fly to match your child's level. A reading tool might offer a little extra practice if they're struggling with fluency. These features can be genuinely helpful — like having a digital tutor who adapts to your child's needs.

But there's a fine line between support and shortcut.

Tools like ChatGPT can be great for explaining tricky ideas — but they can also be used to write entire assignments. And most search engines don't just find information anymore — they summarise it, filter it, and shape it based on invisible preferences and predictions that neither you nor your child set.

A 2024 national survey by Year13 and The Australian found that 75% of high school students are already using generative AI tools for schoolwork. Many said they use it for brainstorming or research — but often without clear guidance on what's okay and what's not. It reflects what a lot of parents and teachers are already seeing: AI is becoming part of how students learn, even if we're all still figuring out the boundaries.

If you're looking for a great way to stay informed about how AI is showing up in classrooms, the AI in Education podcast is well worth a listen. Hosted by educators, it offers real examples, expert insights, and practical conversations about what's working — and what's still up for debate. You can find it at aipodcast.education.

What they see and believe

Platforms like YouTube, TikTok, and Instagram don't just host content — they curate it. The next video your child sees, the next trend they're exposed to, the next idea they start to believe — all of that is shaped by algorithms designed to keep them watching. And while some content can be educational or inspiring, a lot of it isn't. It's loud, fast, attention-grabbing — and it's not there to help them grow, it's there to keep them hooked.

How they're assessed

Some schools and tutoring platforms are starting to experiment with AI that can mark writing, give instant feedback, or even suggest how a student "should" improve. That might save time for teachers, but it also raises big questions. Can a machine understand nuance in a creative story? Can it recognise a student's effort or voice? These tools can be helpful, but they're not neutral — and they don't replace the human judgement and connection that real feedback provides.

What careers they imagine for themselves

As industries change, kids are picking up messages — directly or indirectly — about what jobs are secure, what skills matter, and what paths are "worth it." If AI is replacing humans in certain roles, does that make those jobs irrelevant? Or just different? For a young person still figuring out who they are and what they care about, these shifts can quietly shape their sense of possibility.

And here's the thing:

None of these systems ask your permission.

They don't wait until your child is developmentally ready.

They just appear — and your child adjusts.

Sometimes that adjustment is subtle. Other times, it changes how they think, what they value, or what they believe about their own abilities. As parents, we can't catch everything. But we can start by being aware — and by asking the kinds of questions that keep our kids thinking for themselves.

What's easy isn't always what's good

Many AI tools are designed to make life easier. To speed things up. To reduce friction.

But that very convenience comes with a cost.

When a child uses AI to write an assignment, solve a maths problem, or draft a creative story, the output might look polished. But if they skip the thinking part, they lose something essential: the struggle, reflection, and trial-and-error that real learning is built on.

This isn't just about cheating or laziness. It's about how we grow.

When everything is fast, easy, and neatly packaged, we risk raising kids who are brilliant at prompt engineering but brittle when faced with ambiguity. Who can ace a quiz but can't sit with a tough question. Who expect quick fixes instead of grappling with complexity.

And that's not just an academic concern — it's a developmental one.

In a world of instant answers, we need to actively protect our kids' patience, grit, and self-trust. Moments of discomfort aren't just hurdles — they're where growth happens.

The risk of passive thinking

One of the more subtle — but serious — concerns with AI isn't the fear that it will become too powerful or take over the world.

It's that it might quietly make us stop thinking for ourselves.

When kids grow up in a world where they never have to wait for an answer, where decisions are made for them by an app, and where content is served up based on their every tap or scroll — it becomes really easy to slip into autopilot.

They might not even realise it's happening.

They'll ask a question and accept the first response. They'll follow a recommendation, watch what's next, click what's shown — not because it's the best option, but because it's what's placed in front of them.

And when that pattern repeats over and over?

They might start to believe that AI knows better. That it's smarter. That it sees things more clearly than they do.

So why question it?

That's where the real risk lies.

Not in the technology itself — but in what it can train us to stop doing:

- Asking why.
- Exploring different viewpoints.
- Trusting our instincts.
- Sitting with complexity.
- Making our own decisions — even if they're messy or imperfect.

Children (and adults, if we're honest) can become so used to letting algorithms guide their choices that they forget to drive. And the longer that goes on, the harder it is to notice the quiet hand on the wheel.

This matters far beyond homework or screen time.

It's about identity, agency, and the kind of people our kids are becoming.

Because being thoughtful, curious, and grounded doesn't just help them succeed in school — it helps them become active citizens, capable friends, and resilient humans.

And that's something no AI can ever do for them.

Your role has never mattered more

Here's something worth remembering — especially on the days when parenting feels like shouting into the digital void:

You are still the most important influence in your child's life.

Yes, they might roll their eyes. Yes, their screen might sometimes get more attention than your voice.

But underneath it all, they're still watching how you show up — how you make choices, how you set boundaries, how you respond when the world feels uncertain.

You don't need to have all the answers. In fact, critical thinking isn't built by giving all the right answers — it's built by helping kids navigate uncertainty.

That means asking open-ended questions. Encouraging reflection. Being willing to say "I don't know — let's find out." Helping them see that their voice, their perspective, and their capacity for thoughtfulness still matter — even in an age of machine intelligence.

Because what will set our kids apart won't be their ability to keep up with every new tool. It'll be their ability to stay human: curious, grounded, reflective, and bold enough to chart their own path.

Sometimes your role will be helping them explore AI wisely — showing them how to use it to learn, create, or solve problems.

Other times, it will be encouraging them to unplug — to sit with their own ideas, wrestle with a tough question, or just be a bored, imaginative kid for a while.

Because critical thinking isn't built by giving all the right answers.

It's built by helping kids navigate uncertainty — by showing them that not everything needs to be solved by a screen.

And here's the thing: You're not behind.

You're here, reading, reflecting, thinking ahead — and that matters.

You're in the right place, at the right time, doing the hard and beautiful work of guiding a child through an extraordinary moment in history.

You don't have to have all the answers.

You just need to keep showing up with intention.

Where do we start?

Artificial intelligence isn't a separate topic we can dip into when we have time — it's quietly woven into the fabric of our family lives. It's in the tools our kids use for school, the games they play, the videos they watch, and even the ads we see while scrolling. It's part of the background hum of modern parenting, whether we've noticed it or not.

For many parents, this can feel overwhelming. We wonder if we've missed the memo. If we're already behind.

But take a breath — this isn't about playing catch-up or getting everything right.

This moment isn't about panic or perfection — it's about presence.

It's about choosing to stay alert, curious, and connected.

It's about making space for conversations — even the awkward or uncertain ones.

And it's about remembering that the most powerful tool your child has... is you.

You don't need to have all the answers.

You just need to keep asking good questions, modelling thoughtfulness, and holding onto the values that really matter.

Use the following reflection prompts as an invitation — a way to connect with what matters most in your home, your family's habits, and your child's growth.

Think of this as a gentle pause — a moment to look up and look ahead.

Because navigating the age of AI doesn't start with more screen time or stricter rules.

It starts here, with awareness, conversation, and care.

Questions for thought and conversation

Where is AI already showing up in our family's life — and how do we feel about that?

From streaming platforms to voice assistants to classroom apps, more and more things are being automated. But as we make life easier with technology, it's worth asking — what might we be missing? Things like human connection, real conversation, and knowing when to pause still matter, especially for our kids.

How often do we talk openly with our children about the tools they use?

In a digital world that moves fast and rewards reaction, it's more important than ever to ask thoughtful questions, listen with genuine curiosity, and model healthy digital habits. These aren't just nice-to-haves — they shape how we learn, connect, and lead in increasingly tech-driven environments.

In what ways might AI be making things easier — and what might be lost because of that ease?

Sometimes, the convenience of technology can replace deeper learning — making things quicker, but not always better. When shortcuts take the place of effort, or speed replaces real understanding, kids can miss out on important growth.

What messages is my child receiving about what's valuable — from both humans and machines?

In a world that prizes efficiency and optimisation, creativity, kindness and resilience still need to be celebrated.

What qualities do I want to nurture that no machine can replicate?

Compassion. Imagination. The ability to sit with uncertainty. The courage to question — even when it's inconvenient.

How can I support my child in building a healthy relationship with technology — one that serves them, not the other way around?

Not by rejecting it, but by helping them use it with awareness, boundaries, and purpose.

There's no fixed map for the road ahead.

We can't predict every twist and turn, or know exactly what the world will look like when our kids finish school, start their careers, or raise families of their own. The pace of technological change is fast — and sometimes dizzying.

But that doesn't mean we're lost.

Because even without a perfect map, we still have a compass: our values, our awareness, and our willingness to stay engaged.

Thinking carefully about how AI and other digital systems are already shaping our children's lives gives us something solid to work from. It helps us tune in instead of tuning out — to notice the small moments, the subtle shifts, the habits forming under the surface.

It helps us ask better questions - the kinds of questions that matter. They're how we move from reacting to guiding.

From feeling powerless to being purposeful.

There's no one-size-fits-all answer — and that's okay.

What matters is that we're thinking about it. Talking about it. Staying open, even when it's uncomfortable or complex.

Because the future doesn't just happen to our children.

They grow into it — one decision, one conversation, one guided moment at a time.

And with a little awareness, a lot of heart, and a willingness to keep learning alongside them, we can help them walk into that future not just prepared — but empowered.

Chapter Two: What's Changing in the World of Work

A generation ago, there was a kind of roadmap: finish school, maybe study or train, find a job, build experience, climb the ladder. It wasn't perfect, but it felt familiar. You could generally trust that hard work and time would move you forward.

That's not the world our kids are stepping into.

Work is changing — and it's happening fast. Some jobs are disappearing, others are evolving into something new, and entirely fresh roles are emerging that didn't even exist five years ago — let alone when we were in school. The Australian Government has recognised this shift. Through its National AI Capability Plan, it's working to grow the country's AI workforce, support innovation across industries, and help workers reskill so they're ready for whatever comes next. The goal? To make sure AI benefits all Australians — not just a select few.

One big reason? Automation. Machines and software are now doing jobs that once belonged to people — not just on factory floors, but behind desks, in shops, in hospitals, and even in classrooms. It's not just physical or technical work being affected anymore. It's admin. Retail. Customer service. Data entry. Even parts of teaching, journalism, and law.

And the shift isn't just technical — it's economic. Jobs that once offered a steady income and clear path are being hollowed out. At the same time, demand is growing in areas that rely on something different: creativity, empathy, collaboration, adaptability. The things that make us human.

So what does that mean for our kids?

It means the "safe" paths we might have grown up with aren't necessarily safe anymore. Being employable won't just be about what you know, but how well you can learn, adapt, and bounce back when things change.

That's why this conversation matters now — not just when they're picking subjects in high school.

Let's explore what's changing — and how we can help our kids feel confident and ready for the future ahead.

The disappearing first job

For generations, a first job was more than just a paycheck. It was where teenagers learned to show up on time, deal with tricky customers, work with people they didn't choose, and feel that first flash of independence when they saw their own name on a payslip.

But many of those jobs are changing — or vanishing altogether.

Retail checkout roles are increasingly replaced by self-serve kiosks. Fast food chains are introducing automated ordering systems, some powered by AI that can take drive-through orders faster and more accurately than a teenager under pressure. Call centres are using voice bots. Even simple admin tasks — filing forms, scheduling appointments, responding to emails — are being handed to software.

These roles weren't glamorous, but they were accessible. They didn't require a university degree or years of experience. They gave young people a starting point — and a sense of what it meant to be part of a workplace.

Now, those starting points are fewer and harder to find.

And that has ripple effects. Without these early roles, young people can miss out on key soft skills: communication, conflict resolution, time management, teamwork. Skills that still matter — maybe more than ever — in an age where machines are doing more of the technical heavy lifting.

For parents, this shift raises big questions:

- What will my child's first work experience look like?
- Where will they learn to be responsible, dependable, and adaptable?

- What happens when entry-level jobs don't exist in the same way they used to?

It's not just the income that's disappearing — it's the experience. These jobs were once a rite of passage: learning how to manage time, deal with cranky customers, work in a team, or bounce back after a mistake. These "soft skills" are actually career-critical — and when early job opportunities vanish, so do natural chances to develop them.

How to talk to kids about work that's disappearing

Making space for honest, hopeful conversations

It's not always easy to talk about change — especially when it's tied to something as personal as work. As parents, we want to protect our kids' sense of possibility. But shielding them entirely from what's shifting doesn't help them prepare — or feel empowered.

Instead, we can make room for conversations that are both honest and hopeful.

For younger kids (primary age)

They might hear that "robots are taking over" or that certain jobs are going away. This is a chance to:

- Explain that some jobs are changing, but new ones are being created too.
- Reinforce that who they are matters more than what they become.
- Focus on human strengths: kindness, curiosity, creativity.

You could ask:

"Did you know some shops don't need cashiers anymore? But we still need people to solve problems, help others, and come up with big ideas. That's something machines can't do."

For tweens and teens

This age group is more likely to hear messages about careers being "useless" or futures being unstable. They might feel anxious — or disengaged.

Try:

- Sharing your own path — including any twists, detours, or rethinks.
- Framing work as a mix of skills, relationships, and meaning — not just a paycheck.
- Asking what they enjoy, and where they like solving problems — not just what job they "want."

Helpful phrases:

"The job you do first might not be the one you do forever — and that's okay."

"What you learn about yourself along the way is just as important as what you learn in school."

"Lots of adults are learning new things right now — you're not behind."

These conversations don't need to be perfect. They just need to happen — with curiosity, warmth, and a willingness to sit in uncertainty together.

When kids know we're not expecting them to have it all figured out — and that we're in their corner, no matter how the world shifts — they feel safer exploring who they are and where they want to go.

What jobs are growing

AI and automation might be changing the job landscape, but they're also opening up whole new areas of work — especially in roles that lean on our human strengths, like creativity, empathy, and problem-solving. And it's not just tech companies leading the charge. AI is becoming a major part of how Australia's economy works, popping up in places you might not expect — like finance, farming, hospitality, and even small businesses.

Big picture? There's serious momentum building. PwC estimates around $115 billion could come specifically from gains in productivity — that is, people and businesses working faster, smarter, and more effectively thanks to AI tools. But when you factor in automation more broadly — including things like robotics, smart systems, and other emerging technologies — government projections put the potential impact closer to $600 billion a year. The key takeaway: AI is changing how we work and what kinds of jobs are out there.

To keep up, Australia's building its own tech smarts — growing its local AI talent, investing in homegrown innovation, and making sure we're not relying too much on overseas tools. Back in 2014, there were just 800 people working in AI here. By 2023, it had jumped to over 33,000 — and by 2030, it could hit 200,000. That kind of growth means our kids will be stepping into a job market that looks very different from the one we knew.

Here's where the growth is happening — and why it matters for our kids:

Healthcare and aged care

An ageing population and rising demand for mental health support are fuelling enormous growth in healthcare. We're seeing consistent need for:

- Registered nurses
- Aged care workers
- Allied health professionals (like physios and occupational therapists)
- Mental health practitioners

Healthcare isn't just a "safe" career choice — it's one that's becoming more important as the population ages and community care becomes more complex. *Source: BSN Australia – Top 5 Healthcare Jobs Shaping Australia's Future in 2025*

Education and training

Australia is facing a national teacher shortage — but that's only part of the story. Demand is also rising for:

- Early childhood educators
- Vocational education and training (VET) instructors
- University lecturers and tutors
- Education support workers

As governments invest in lifelong learning and skills training, education continues to be a vital growth area — especially for those who can engage and support diverse learners. *Source IBA Partner – Australia's Growing Job Demand (2025)*

Renewable energy and infrastructure

As the climate crisis drives urgent investment in sustainability, we're seeing strong demand across:

- Renewable energy (solar, wind, hydrogen)
- Smart city infrastructure
- Sustainable building and retrofitting

Jobs are growing for civil engineers, project managers, electricians, and skilled tradespeople — especially those with training in green technologies. *Source: IBA Partner – Future-Focused Jobs in Sustainability 2024*

Technology and AI-related roles

While AI is disrupting some industries, it's also fuelling demand for:

- Software developers
- Cybersecurity analysts
- Data scientists and engineers
- AI ethics consultants and prompt engineers

These roles don't just need technical know-how — they demand human-centred thinking. Skills like ethical reasoning, cultural awareness, and emotional intelligence are increasingly essential. Designing a health app? You'll need to consider bias in the data. Building a chatbot? You'll need empathy to get the tone right. It's tech — with a human edge. *Source: PwC Australia – AI Jobs Barometer Report (2025)*

Skilled trades and technical roles

Despite the rise of automation, skilled trades are still in high demand — especially in sectors supporting infrastructure, manufacturing, and smart technologies. This includes:

- Plumbers
- Electricians
- Mechanics
- HVAC (Heating, Ventilation, Air Conditioning) technicians
- Building and construction workers

These jobs are harder to automate and deeply embedded in local economies — making them stable and essential. They also require manual dexterity, and real-time response to unpredictable situations, making them harder to replace with AI alone. However, AI can greatly assist in areas such as predictive maintenance, scheduling, or data analysis. *Source: IBA Partner – Australia's In-Demand Industries (2025)*

Mental health and human services

There's also rising demand in professions focused on wellbeing, support, and care — areas where empathy, emotional insight, and human connection are irreplaceable. These include:

- Psychologists and counsellors
- Youth and social workers
- Wellness coaches and community outreach roles

As awareness of mental health grows, these careers are becoming not just valued, but vital. *Source: BSN Australia – Mental Health Careers on the Rise*

The jobs of the future aren't only technical. Yes, digital and data skills matter — but so do communication, collaboration, empathy, and ethical thinking.

These growing sectors suggest something hopeful: that even in a high-tech future, our most human qualities will still be in high demand.

“What’s a prompt engineer?” — jobs that didn’t exist 5 years ago

A quick glimpse at how fast the world of work is evolving

Tech isn’t just replacing roles — it’s creating entirely new ones. These are real job titles being advertised or filled in Australia and globally. Most didn’t exist when today’s high schoolers were in primary school:

Prompt engineer

Helps AI systems like ChatGPT generate better, more accurate responses — by crafting clear, detailed instructions.

Ai ethicist

Guides organisations on how to use AI fairly and responsibly — asking questions about privacy, bias, and impact.

Virtual production designer

Combines film, gaming, and AI tools to create realistic sets and visual effects without leaving a studio.

Cybersecurity behaviour analyst

Studies how people interact with systems to prevent data breaches — blending tech with psychology.

Drone traffic manager

Oversees the safe routing of drones in urban areas — yes, that’s now a thing.

The takeaway? Many of the jobs our kids will have don’t exist yet — and the best prep is to help them stay curious, adaptable, and open to learning.

The human skills machines can't replace

As technology continues to race ahead, it's easy to feel like the best thing we can do for our kids is help them "keep up." We hear a lot about coding clubs, robotics kits, and STEM pathways — and those things are valuable, no question. But in all the buzz around tech, it's easy to overlook something just as important: the quiet, steady human skills that machines still can't touch.

Skills like kindness. Integrity. Creative thinking. Knowing how to listen, how to apologise, how to ask thoughtful questions, or how to keep going when something feels hard or uncertain.

These aren't just nice-to-haves. They're the foundation for being adaptable, employable, and able to thrive — not just survive — in a fast-changing world.

Some people call them "soft skills," others call them "durable" or "human" skills. Whatever name they go by, they're the things that help someone not just do a job, but work well with others. To lead. To reflect. To make good choices. And to bounce back when things don't go to plan.

And in a world where machines can increasingly handle the technical side of things? These human qualities matter more than ever.

According to the World Economic Forum Future of Jobs Report 2023, globally, almost 40% of workers' current skillsets are expected to be outdated or transformed by 2030. While AI and data skills are in demand, employers are consistently saying that what they need most are human strengths: analytical and critical thinking, resilience, adaptability, emotional intelligence.

Because yes — AI can process data, write reports, and optimise systems. But it can't build trust. It can't show empathy. It can't guide a team through a crisis or spark an idea that's never existed before. Those are the things only people can do — and they're what will help our kids thrive, no matter how the world of work evolves.

In fact, Australia's Voluntary AI Safety Standard reinforces this idea – that no matter how advanced technology becomes, it still needs human oversight. The standard outlines 10 guardrails for responsible AI use, including principles like "Act Fairly" and "Know Your Audience." These aren't just technical checkboxes – they're reminders that AI systems should reflect our values, and that things like empathy, judgment, and care still need to come from us. We'll explore these standards more in Chapter 6, where we take a closer look at how Australia is responding to the rise of AI.

So while it's great to support your child's interest in tech, it's just as important to nurture their ability to think deeply, relate to others, and stay grounded in who they are. In a future full of smart machines, those are the skills that will set them apart.

What are these skills – and why do they matter?

Let's break them down.

Critical thinking

Being able to ask why, even when an answer is already given. This matters in a world where information is everywhere – but not all of it is accurate or fair. It means teaching kids not to blindly trust the first result, or the loudest voice, or even the chatbot that sounds confident.

Creativity

The ability to come up with something original – a story, a new approach, an unexpected solution. AI is great at remixing what already exists. But imagining something that doesn't yet? That's still a human superpower.

Emotional intelligence

Understanding feelings – your own and other people's. Reading the room. Navigating friendships, teamwork, conflict. Machines can mimic empathy, but they don't feel it. That's where people shine.

Curiosity

Asking questions. Exploring for the sake of learning, not just getting a right answer. AI gives information fast — but curiosity is about going deeper, and not being afraid to say, "I don't know, but I want to find out."

Ethical judgement

Doing the right thing — even when it's not the easy thing. Knowing that just because something can be done, doesn't mean it should be. In a world of fast answers and fast tech, kids need space to develop a moral compass.

Adaptability

Being flexible when things change — because they will. Whether it's a new app, a new job, or a global crisis, the ability to shift gears and stay steady is gold.

Communication

Not just typing or talking, but expressing ideas clearly, listening deeply, and understanding the nuance of human interaction. Whether it's writing a thoughtful email or resolving a misunderstanding, communication is where leadership begins.

These skills start at home

You don't need to be an expert to nurture these qualities. In fact, you're probably already doing it — more than you realise.

When you:

- Encourage your child to talk through a problem instead of solving it for them
- Ask "what do you think?" before giving your opinion
- Help them bounce back after disappointment
- Get curious about a question together instead of rushing to Google
- Model kindness, fairness, and reflection in your own choices

... you're already building the kind of skills that last a lifetime.

These qualities don't need a coding bootcamp. They need time, space, and encouragement. They're built in conversation, in play, in moments of frustration and joy. They grow when kids feel safe to explore — and when they see those values modelled by the adults around them.

What this means going forward

We can't predict exactly what jobs will exist ten or twenty years from now — and honestly, neither can the experts. But one thing is clear: the only constant will be change.

Tools will keep evolving. Industries will keep shifting. And the ability to stay thoughtful, curious, kind, and grounded? That's what will help your child navigate whatever comes next — not just as a worker, but as a whole human being

Yes, let's help them get comfortable with technology. Let's show them how to use it responsibly and creatively. But just as importantly, let's teach them how to stay human — how to think critically, relate to others, ask big questions, and adapt with grace when things don't go to plan.

Because no matter how clever machines become, the most powerful thing your child can bring to the table... is themselves. Is it cheating, or is it just changing?

That's especially true as the pace of change starts to outgrow the systems we've traditionally relied on to prepare young people for work. According to the World Economic Forum's Future of Jobs Report (2023), nearly 40% of workers' core skills are expected to shift by 2030. And many employers say what students learn in their first year of university may already be outdated by the time they graduate.

That's why 85% of global companies plan to prioritise upskilling and reskilling their workforce by the end of the decade — not just in tech roles, but across all sectors. *Source: World Economic Forum, 2023*

The message is clear: learning can't stop at school. It has to be lifelong, flexible, and increasingly focused on how to think — not just what to memorise. Through the National AI Centre and AI Adopt Centres, small businesses and everyday workers are gaining access to upskilling tools and real-world support. The government is also funding initiatives to ensure AI literacy is accessible and inclusive — part of its broader commitment to fair and safe AI for all.

As parents, we can support that shift by modelling curiosity, encouraging reflection, and helping our kids feel confident learning new things — even when it's uncomfortable. Because in the end, it's not about keeping up with the machines. It's about helping our kids stay grounded in the kind of skills and values no machine can replicate.

That's how we raise children who don't just survive a changing world — they help shape it.

What this means for the economy — and our kids' futures

It's one thing to say jobs are changing. It's another to look at how that shift affects the world our kids are growing into — and what that might mean for their lives, choices, and stability in the decades ahead.

Because this isn't just a work problem.

It's a ripple effect that touches everything from housing and education to mental health, lifestyle expectations, and social mobility.

Let's unpack what's already shifting — and why it matters to us as parents.

The middle-class squeeze is real

For a long time, the middle class felt like a safety net. If your child finished school, worked hard, got a degree, and found a decent job — they'd likely be okay. They could buy a home, support a family, and look forward to retirement.

But that path is changing.

Jobs that used to offer reliable, well-paid work — like admin roles, call centres, retail management, and even junior legal or finance jobs — are now being automated or moved offshore. The Grattan Institute says that middle-skill jobs are the most at risk, while growth is happening mostly in lower-paid or highly specialised roles.

This "hollowing out" of the middle isn't just an economic issue — it's a parenting one. It means our kids might face more financial stress, more competition, and more pressure to stand out — even if they've done everything "right".

The ripple effect: when middle-income jobs disappear

When we hear about jobs being lost to automation, we often picture certain industries — like retail, transport, or office admin. And yes, those areas are changing fast. But the ripple effects go much further than we might think.

It's not just about one person losing a job. It's about what happens when thousands of people lose steady, middle-income work at the same time. Because those jobs do more than support individual families — they help keep the whole economy moving.

Think about all the everyday spending that flows from stable work: family dinners out on a Friday night, school sports fees, weekend haircuts, tutoring, local holidays. When people have a bit of breathing room in their budget, they support businesses across many different sectors.

But when those jobs disappear — or the wages don't keep up — that spending slows down. And it doesn't just affect the industries you'd expect. It can hit hospitality, events, tourism, the arts, and even things like education support services. Fewer people spending means fewer jobs — even in areas that seem totally unrelated to tech.

So while AI might not be replacing your job or your child's future job directly, it still has the power to shift the world they're growing up in.

That's why this isn't just a tech story. It's an economic one. And for parents, it's a reminder that helping our kids prepare for the future isn't just about getting them to learn new tools. It's about building the kind of flexible, human skills that will help them adapt — no matter how the world of work changes.

The cost of living doesn't match the new normal

While some jobs are disappearing or changing fast, the cost of living hasn't exactly eased up to match. If anything, it's getting harder for young people to find their financial footing.

Housing is expensive — not just buying, but renting too. Tertiary education comes with a hefty price tag, and even the "Plan B" of starting your own business isn't simple anymore. Launching something on your own might sound empowering, but it now involves navigating everything from digital tools and social media algorithms to competing with AI-generated content. It's not the side-hustle dream it's often made out to be.

All of this can leave young people stuck in a tough spot. They may feel like they're doing all the right things — studying hard, applying for jobs, building skills — and still struggling to keep up. Many end up needing to live at home longer, juggle multiple gigs, or retrain again and again just to stay afloat.

And it's not just a financial issue. When it feels like progress is always just out of reach, it can start to chip away at confidence and independence. It can make young people doubt their own choices or feel behind their peers. For some, it affects their mental health, creating a quiet pressure that parents may not always see — but that many young people carry.

As parents, we can't fix the economy. But we can create space to talk openly about these pressures, remind our kids that their worth isn't tied to a job title, and help them build the kind of self-belief and resilience that lasts longer than any trend.

Gig work isn't a safety net — it's a rollercoaster

The rise of gig and contract work is often marketed as "flexibility" — and sometimes, it really is. For some people, being able to choose your hours or work across different projects can be empowering. But for many young workers, that flexibility comes with a trade-off: no superannuation, no paid leave, and no real job security.

In fact, according to the McKell Institute, Australia's workplace laws haven't kept pace with this shift. Gig workers — whether they're driving for a rideshare company, delivering food, or freelancing online — are often left without the protections that traditional employees take for granted. That means they're more vulnerable to financial stress, sudden income loss, and burnout.

If your child wants to build a future as a creator, a freelancer, or their own boss — that's something to be proud of. But talent alone isn't enough. They'll also need a solid understanding of how digital platforms work, how to protect their rights, and how to manage money in an unpredictable income world.

That means learning about things like tax, super, budgeting, pricing their time, and planning for the slow months — things many schools still don't teach in a meaningful way.

This is where the future of work is headed — and it's reshaping what "success" looks like.

Inequality is widening

As AI tools become more powerful, the biggest benefits often go to the kids who already have a head start — access to good devices, fast internet, private tutoring, or simply parents who have the time and knowledge to help them figure it all out.

But not every family has those resources. In fact, the Australian Digital Inclusion Index shows that children in remote areas, low-income households, and many Indigenous communities still face real barriers when it comes to accessing technology — let alone using it with confidence.

And here's where AI can quietly widen the gap. If AI is shaping how we learn and work, then missing out on those tools — or only using them in limited ways — can affect a child's confidence, their curiosity, and their future opportunities.

Some kids are learning how to use AI to spark new ideas or explore complex questions. Others may only encounter it as a basic homework shortcut — or not at all. That difference matters. Because the longer we wait to close the gap, the harder it becomes to catch up.

This isn't just a tech issue. It's a question of equity, access, and inclusion — and it's something we need to keep front of mind in both policy and parenting.

So what does this mean for our kids?

It means the old advice — "just work hard and get a good job" — doesn't quite cut it anymore.

Of course, effort still matters. But in a world where industries can shift overnight and job titles don't mean what they used to, kids need more than just determination. They need adaptability. Creativity. The ability to ask good questions, spot patterns, and make decisions in the grey areas. They need a strong sense of who they are — and where they want to go — even when the path isn't clear.

And maybe most of all, they need support. Someone in their corner who can help them make sense of it all when the future feels uncertain or overwhelming.

That's where we come in.

As parents, we can't control the economy, the job market, or the pace of change. But we can give our kids something just as powerful: the confidence to keep learning, the tools to stay curious, and the reminder that their value goes far beyond any job title or Australian Teriary Admission Rank (ATAR).

The world might not come with the same guarantees it once did — but we still have the chance to help our kids build a future they can feel proud of. One that they can shape, grow into, and thrive in.

Why this matters

The world of work is changing — fast. Jobs we once considered stable are being reshaped or replaced, while new roles emerge that didn't exist even a few years ago. For our kids, this isn't a future problem. It's the reality they're growing up with right now.

What's disappearing isn't just tasks, but key stepping stones — like casual after-school jobs that helped build confidence, taught responsibility, and eased young people into working life. These early experiences weren't just about money; they helped kids learn how to show up, solve problems, and work with others.

AI and automation bring exciting new possibilities — but they also challenge us to think differently about what skills really matter. Because what machines do well (speed, pattern recognition, data crunching) isn't what makes someone a thoughtful teammate, a caring leader, or a creative problem-solver.

That's where human strengths come in. Qualities like empathy, adaptability, creativity, and emotional insight — the things AI can't replicate — will only become more important as our kids step into their own futures.

It's important to stay tuned in to how the world is shifting — and help our kids build the kind of inner toolkit that keeps them steady, curious, and ready to grow with it.

Questions for thought and conversation

Take a moment to reflect or chat with your child about the world of work they're heading into. You don't need to have all the answers — just asking the questions together can make a big difference.

What did "work" look like when I was growing up — and how is it already different today?

Are there jobs that don't exist anymore, or new ones I didn't imagine as a child?

What messages is my child hearing about success and security?

Are they learning that a degree guarantees stability — or that careers are more flexible (and unpredictable) now?

What kinds of work feel meaningful to me — and how do I talk about that at home?

Do I share what I value in my own work life? How do I talk about purpose, creativity, or collaboration?

How can I help my child see work as more than just a job title?

Are we exploring their interests, strengths, and the human skills that will help them adapt — no matter what the job market looks like?

Am I encouraging problem-solving, initiative, and communication — or mostly focusing on grades and tasks?

Where do we create space for growth, experimentation, or making mistakes?

Work is changing — and with it, the expectations, pressures, and possibilities our kids will face. The paths that once felt stable are shifting, and the stepping stones we took for granted — like a first job at the local shop — are harder to find.

But while the world of work may look different, our role as parents hasn't changed as much as we think.

We're still the ones helping our kids build the habits, values, and self-belief they'll carry into whatever comes next. That might mean having honest conversations about uncertainty. Helping them practise problem-solving when plans fall through. Or reminding them that success isn't just about job titles — it's about character, effort, and growth.

The future of work may not come with clear instructions. But with curiosity, empathy, and a steady presence in their corner, we can help our kids shape futures that feel meaningful — even in the midst of change.

Chapter Three: AI in Education

How AI is showing up in classrooms — and what our kids really need to learn

In Chapter 1, we touched on some early examples of how AI is being used in Australian classrooms — from chatbots like EdChat to tools like Education Perfect, and that big rollout by Brisbane Catholic Education. But education deserves more than just a mention.

Because when AI shows up in schools, it's not just about smart tools — it's about how we protect space for deep thinking, human connection, and the kind of learning that lasts. When we think about school, many of us picture the same staples: teachers at the front of the room, exercise books, timetables, and homework. But today's classrooms are beginning to look — and feel — very different.

Artificial intelligence isn't some far-off concept for our kids. It's already here. Not as robots teaching the class, but as smart tools helping students learn, teachers personalise lessons, and schools manage their day-to-day tasks. In some cases, kids might not even realise they're using AI — but it's shaping their experience all the same.

Platforms like EdChat (SA), NSWEduChat, Mathspace, and Education Perfect are already in use across both public and private schools in Australia. From personalised assignments to analytics that help track student progress, AI is powering parts of the classroom. Schools in NSW, SA, QLD, and elsewhere have been piloting these tools—driven by the national educational push to use AI in ways that benefit learning and safety.

Fast, convenient—and sometimes confusing. That's why the government released the Australian Framework for Generative AI in Schools in late

2023. It sets out six principles—like fairness, privacy, transparency, and accountability—to guide how AI is used in education (for example, making sure students' data isn't shared or sold) .

This isn't about fear or resistance. AI can be a powerful tool in education—but when it's introduced this quickly, questions and concerns naturally arise.

- How do these tools impact our kids' learning?
- Are teachers supported to use AI well?
- Is our child's privacy being respected?
- And how do schools ensure that the human relationship—the conversations, creativity, critical thinking—remains central?

Parents don't need to be tech experts. But understanding how AI is being woven into school life—and what guardrails are in place—helps you step confidently into these new conversations with your child and their teachers.

Let's take a closer look.

What tools are already in use?

When we think about AI in schools, it's easy to imagine robots at the front of the classroom — but that's not what's actually happening. Most of the AI already in use is working behind the scenes, helping personalise learning, reduce teacher admin, and support students in real time. And these tools aren't just in high-tech private schools. They're being used in public classrooms across Australia right now.

Here are some examples you might already have seen (or will soon):

EdChat – like a homework helper that doesn't just give answers

- This Aussie-designed chatbot can explain tricky topics in plain language.

- Kids can ask it things like, “What’s a simile?” or “Why do volcanoes erupt?” and get help that lines up with what they’re learning in class.
- It’s used in some public schools already and helps build confidence without doing the work for them.

NSWEduChat - a safe, curriculum-aligned AI assistant for nsw classrooms

- This chatbot supports students by explaining tough topics in clear, curriculum-aligned language.
- It’s designed to promote understanding, not just fast answers — helping kids think through problems.
- Rolled out late 2025 across all Year 5 to Year 12 students in NSW public schools, it reflects the Department’s focus on safe, educational use of AI. *Source: nsw.gov.au Sept 2025*
- Teachers can review student interactions, supporting transparency and guided learning.

Mathspace – A digital math tutor that gives real-time feedback

- Used in many Australian schools, Mathspace adapts to each student’s skill level.
- If a child makes a mistake, it gives step-by-step guidance to help them understand why — not just mark it wrong.
- It’s like having a patient tutor who’s always available.

Education Perfect – Interactive learning across subjects

- This platform is used in over 3,000 schools, covering subjects like science, English, and languages.
- It includes lessons, quizzes, speaking tasks, and games — all powered by AI.
- For example, students learning Japanese can practise speaking and get instant feedback on their pronunciation.

Corella AI– Queensland's safe AI platform for schools

- Recently launched by the Queensland Department of Education, Corella AI is a custom-built tool giving state school teachers and students safe access to AI.
- Unlike open platforms such as ChatGPT, Corella is designed specifically for classrooms: no advertising, no data harvesting, and content aligned with the Australian Curriculum.
- Teachers can use it to help with lesson planning or marking, while students can use it for personalised learning support — all within a monitored and secure environment.

The goal is to let students benefit from AI while ensuring safety, privacy, and trust are built in from the start.

Teacher support tools – saving time so teachers can focus on kids

Some schools are now trialling tools that help with lesson planning and report writing — not to replace teachers, but to ease the load of repetitive admin. Platforms like Kuraplan and Eduaide.AI offer guided templates aligned to the Australian curriculum, helping teachers quickly generate lessons and assessments.

Meanwhile, tools like Report Wizard are being used to draft personalised report comments, giving teachers a head start without removing their professional judgement. And in Queensland, Brisbane Catholic Education is developing its own AI assistant using Microsoft technology — part of a broader effort to support more than 140 schools with time-saving tools.

These initiatives are still evolving, but the aim is clear: to free up teachers' time so they can focus more on what really matters — connecting with students.

Why this matters for families

When teachers spend less time buried in admin, they have more energy for the things that really support learning — noticing when a child is struggling, planning creative lessons, or having those encouraging one-on-one chats that make a difference. AI might be behind the scenes, but its impact shows up in the everyday moments that shape how kids feel about school.

AI might be working behind the scenes, but its impact shows up in the everyday moments that shape how kids feel about school.

And it's not just individual schools making changes — even the people who shape national education policy, like ACARA, are exploring how to include AI in classroom learning in thoughtful, age-appropriate ways.

At the same time, some of the most powerful changes are happening behind the scenes — in the systems and infrastructure that support schools every day.

For example:

- Some tools help flag student well-being concerns based on patterns in attendance, engagement, or assessment data.
- Others analyse classroom performance to personalise learning plans or identify students who may need extra support.
- There are platforms that support teachers with smart timetabling, draft reports, or communication tools — saving hours of admin time.

Often, these tools don't even look like "AI." They're just part of the platforms schools already use. But they still influence learning decisions, time allocation, and student support — which makes it important for parents to stay curious and ask questions like:

- What tools does our school use to track learning or wellbeing?
- How is student data protected?

- Are teachers supported in understanding how these tools work?

So while robots aren't taking over the classroom, AI is already part of how schools operate — not in the future, but right now.

Stats & insights: AI in Australian schools

80% of Australian secondary schools are already using AI tools in teaching and administration. 1 in 5 schools plan to expand their AI programs further in 2025. *Source: Education Daily, 2025*

In South Australia, AI chatbots used for English language assessment have cut testing time from 30 minutes to just 52 seconds — saving around 15,000 hours of teacher time. *Source: The Australian, 2024*

A 10-week study involving 20,000 students across Australia and New Zealand found that AI-powered feedback tools led to a 47% improvement in the quality of students' final written responses. *Source: Adelaide Now, 2024*

65% of Australian teachers report using AI in academic work, such as lesson planning, resource curation, or grading. *Source: Open2Study, 2025*

The Australian Framework for Generative AI in Schools was endorsed in mid-2025 to guide the safe, responsible use of AI in all education sectors. *Source: Department of Education, 2025*

Over 3,000 Australian schools use Education Perfect, and tools like Mathspace and EdChat are being piloted in both public and private classrooms. *Source: Education Perfect, 2024*

Ways to stay engaged

AI is already part of your child's learning — not in a flashy or scary way, but in small, helpful tools designed to support teachers and students.

Regardless of your level of tech know-how, it's important to stay involved. Just being aware of these tools can help you:

- Ask better questions at parent-teacher interviews.
- Support your child with homework (and understand the tools they're using).
- Feel confident that education isn't being handed over to machines — it's still very human.

Talking about AI in schoolwork

Has your child mentioned using apps or digital tools at school?

Next time you're chatting, try asking:

- "What's one digital tool you've used lately that helped you understand something better?"
- "What made you decide to use that app or AI tool?"
- "Was there a part of the task that you still had to think through yourself?"
- "If you had to explain it to someone else without using the tool, how would you do it?"
- "Do you think it helped you learn more — or just get it done faster?"

You might be surprised by how much AI is already part of their day — in small but meaningful ways.

A note on tertiary education

While this chapter focuses on school-aged learners, it's worth noting that AI is also transforming Australia's universities and TAFEs. From AI-assisted research tools to virtual tutors and automated marking systems, higher education institutions are exploring ways to use AI to support both teaching and learning.

Students in tertiary settings are often using tools like ChatGPT, Grammarly, and discipline-specific platforms to draft essays, summarise readings, or get unstuck on assignments. These tools can be helpful – but they also raise important questions about originality, academic integrity, and digital literacy.

Many universities are now updating their policies to reflect the rise of generative AI – not to ban it outright, but to guide students in using it responsibly. The emphasis is shifting from detection to education: helping students understand when AI can assist their learning, and when it crosses a line.

AI detectors: why they miss the point

As AI tools like ChatGPT and other writing assistants become more common in schools, many systems have tried to respond by using AI detectors – programs that claim to tell whether a piece of writing was created by a student or by a chatbot.

It sounds simple enough, but here's the problem: these detectors just don't work reliably.

They often flag genuine student work as "AI-generated," especially if it's written clearly, simply, or by students learning English as an additional language. On the other hand, they can completely miss text that's actually been written or heavily edited by AI.

Even the companies that build AI detectors warn against using them for punishment. The risk of false accusations and the stress that causes is real.

Instead of focusing on catching AI use, the more useful approach is to focus on guiding it.

That means helping students understand when and how it's okay to use AI — for brainstorming, feedback, or learning support — and when they need to do the thinking themselves.

Schools are starting to shift from "AI detection" to "AI literacy": teaching students how to use these tools responsibly, reflectively, and transparently. Because the real goal isn't to stop them from using AI — it's to help them use it well.

The big opportunities — and big questions

There's no doubt AI has the potential to make learning better. Smarter. More personal. But like any new tool in our kids' lives, it comes with trade-offs — and questions worth asking.

Let's look at both sides: the exciting possibilities, and the things parents should keep an eye on.

What AI can do well

AI can support teachers and learners in powerful ways. When used thoughtfully, it can help tailor education to each child's needs — rather than relying on a one-size-fits-all approach.

Here are some of the biggest upsides:

- More personalised learning

  Tools like Mathspace or Education Perfect adjust in real-time to a student's answers, offering extra support where they're struggling — or new challenges when they're ready to move ahead.

- Support for neurodiverse learners and ESL (English as a Second Language) students

AI tools can provide captions, translations, read-aloud features, and simplified instructions — all of which can help kids learn in ways that suit them best.

- Less admin for teachers

With AI helping to draft reports, sort through data, or plan lessons, teachers can spend more time doing what they love: actually teaching.

- Spotting challenges early

Some tools help flag learning gaps or behavioural trends earlier than a human might notice — making it easier to offer timely support before problems grow.

These benefits aren't just theoretical — they're already being trialled in classrooms across Australia. And when AI is used to enhance teaching (not replace it), it can help every child feel more seen and supported.

What parents should be asking

At the same time, there are important conversations to be having — especially as AI tools become more embedded in everyday school life.

Here are a few things that deserve attention:

- Who's collecting the data — and what happens to it?

Many AI tools collect student data to function properly. But where is that data stored? Who can access it? And how is it protected? Parents have a right to understand this.

- Is the AI fair and inclusive?

AI systems learn from past data — and if that data includes bias (even unintentionally), it can affect how the tool performs for different students. We want to be sure all kids, from all backgrounds, are treated equitably.

- Are kids learning to think — or just to prompt?

With tools like ChatGPT available, it's easy to find fast answers. But are students learning how to think, ask great questions, and form their own ideas — or just how to phrase a good prompt?

- Could teachers lose essential skills over time?

If AI takes over planning, grading, and feedback, what happens to teachers' professional judgement and creativity? Are we unintentionally deskilling the very people we trust to shape our kids' learning?

What AI can't do well (yet)

Let's be clear: AI is a powerful tool. But it's not a teacher.

It can generate practice questions, explain tricky concepts, or offer quick feedback. But it can't read the room. It can't sense when a student is discouraged, disengaged, or deeply proud of a small win. It can't model curiosity or share real-life stories that make a topic stick.

In short? AI can support learning — but it can't replace the warmth, wisdom, and human insight that educators and families bring.

Here's where it still falls short:

- It doesn't build real relationships.

A chatbot might help with a maths problem, but it can't notice when your child looks tired, anxious, or quietly proud of their effort. Teachers connect in ways machines simply can't.

- It can't understand context.

AI is trained on patterns — not meaning. It might give a technically correct answer that completely misses nuance, emotion, or cultural sensitivity.

- It's not always right.

AI tools like ChatGPT can sound confident — even when they're completely wrong. Kids (and adults!) still need critical thinking to know when to pause and ask, "Does this actually make sense?"

- It can't model human values.

Respect. Fairness. Courage. Patience. These aren't downloaded — they're developed through human example. Children don't just learn what we teach; they learn how we live.

So yes, AI can be helpful. It can explain, reinforce, and personalise. But what it can't do is teach what it means to be human — and that's still the most important lesson of all.

Skills that matter more than ever

If AI is changing how our kids learn, then we need to take a closer look at what they're learning — and why.

In a world where technology can instantly provide facts, summaries, and answers, the real value lies in the human skills that AI can't copy. These aren't just "nice to have" extras — they're the foundation for how young people will thrive in work, relationships, and life.

So, what skills will actually matter in an AI-rich future?

Here's what the experts — and everyday experience — are showing us:

Critical thinking

It's not about just accepting the first answer (especially when it's written by a confident-sounding chatbot). Kids need to learn how to:

- Question what they read and hear

- Compare different viewpoints
- Spot when something might be misleading or biased

Example: If your child uses an AI tool to research a science project, critical thinking helps them decide which facts are reliable — and which ones might need double-checking.

Creativity

AI can remix and replicate — but it doesn't imagine. Inventing something new, telling a fresh story, or designing a better solution? Still very human.

- Problem-solving through art, writing, or innovation
- Thinking outside the box
- Connecting ideas in unexpected ways

Example: A student might use AI to generate ideas — but creativity helps them decide which ideas matter, and how to make them their own.

Communication

As information gets faster, clarity becomes more valuable. So does the ability to listen well and connect with others.

- Explaining complex ideas simply
- Listening actively
- Navigating tough conversations

Example: Whether it's a group assignment or a part-time job, good communication builds trust and helps kids be heard and understood.

Problem-solving

The future will bring plenty of challenges — from climate change to shifting job markets. Kids need to get comfortable:

- Trying different approaches
- Making mistakes — and learning from them
- Working through tricky problems without giving up

Example: An app can help practise maths facts, but problem-solving is what helps a child figure out how to apply that knowledge in a real-world scenario.

Adaptability and resilience

With tools, platforms, and expectations changing all the time, being flexible and emotionally steady is more important than ever.

- Coping when things go wrong
- Staying curious in uncertain times
- Learning how to regroup and try again

Example: A new AI learning tool might feel overwhelming at first. Resilience helps kids keep going — even when it's not easy.

Ethical thinking

The old adage, just because you can do something doesn't always mean you should, applies to AI as much as to kids! They will need a strong internal compass to navigate questions about fairness, truth, and responsibility.

- Understanding consequences
- Asking, "Who does this help — and who might it harm?"
- Making values-driven choices

Example: An AI tool may suggest ways to write an essay faster — but ethical thinking helps your child decide what's okay to use, and what's crossing a line.

Digital literacy

More than just knowing how to use a device, this means understanding the systems behind it.

- How does AI work?
- What is it doing with your data?
- Can it be biased — and how would you know?

Example: A child who knows how to spot bias in an AI tool is better equipped to challenge unfair systems — not just accept them.

Quick fact:

The World Economic Forum estimates that over 40% of core skills for students today will have changed by 2030. That's less than a decade away — and a strong signal that traditional learning alone won't be enough. *Source Word Economic Forum Future of Jobs Report 2025*

These aren't skills you'll always find on a report card. But they're the ones that will matter most — at school, at work, and in life.

And the good news? These skills start at home. In dinner table conversations, in games, in everyday decisions. Every time you ask your child what they think, let them try something new, or talk through a tricky situation together — you're building their future-ready toolkit.

Australia's approach — and where it's headed

When it comes to AI in schools, Australia isn't sitting still. While we'll explore the broader national response to AI later in this book, it's worth knowing what's already happening in the education space — because it's more than just talk.

Across public and private schools, governments are working to strike a balance between innovation and care. They're asking: How can we use these powerful new tools to support teachers, improve learning, and keep students safe?

Here's what's already underway — and why it matters for families.

National guidance for schools

In 2023, Australia released the Framework for Generative AI in Schools, giving schools a roadmap for using tools like ChatGPT, AI tutoring assistants, and learning platforms responsibly. It covers everything from student safety and data privacy to how AI can be used without undermining core teaching values.

A 2024 review of the framework emphasised the same message: AI should assist learning, not replace it. Teachers should remain central to decision-making, and any AI use must support equity, inclusion, and critical thinking — not shortcuts or surveillance.

Ethical foundations and guardrails

Australia is also taking steps to make sure AI in schools aligns with our shared values.

- The AI Ethics Principles developed by the Australian Government include things like "Do no harm," "Benefit people," and "Know your audience." These aren't just nice ideas — they're designed to guide how AI is used across sectors, including education.
- In response to public feedback, the government is now developing mandatory guardrails to back up these principles with clear protections — especially when it comes to data, fairness, and accountability in learning environments.

This kind of ethical scaffolding helps build public trust — something that's especially important when it comes to children and young people.

Building skills for teachers and students

As part of the National AI Capability Plan, the government is investing in building digital and AI literacy — not just for kids, but for educators too. Because let's face it: it's hard to teach what you've never been trained in.

- Programs like the National AI Centre and AI Adopt Centres are giving schools and small businesses practical, on-the-ground support. Think: hands-on training, simple resources, and expert advice — not just high-level policy.

This work helps ensure teachers aren't left to figure it out on their own — and that students learn not just how to use AI, but how to use it wisely.

How Australia compares globally

Australia is making steady progress, but the global pace is quick. Other countries offer interesting models:

- Finland has become a leader in public AI literacy, offering free online training to citizens of all ages, including students and their families.
- Singapore is integrating AI across its entire school system — from coding basics to ethical use.
- The UK is embedding AI education into digital citizenship, encouraging students to think critically about the tools they use.

Australia's approach so far has been thoughtful and collaborative — but as the tech accelerates, so too must our efforts to stay inclusive, ethical, and future-ready.

"Australia's message is clear: AI can support learning — but it should never replace the human heart of education."

In 2022–23, nearly $500 million worth of AI, computer vision, and machine learning projects were supported through Australia's Research and Development Tax Incentive — helping drive innovation in sectors including education. (Source: National AI Capability Plan)

What parents can do

You don't need a tech degree to help your child thrive in an AI-enhanced classroom. The most powerful thing you can offer isn't technical knowledge — it's your presence, your questions, and your perspective. Kids today are growing up with tools we never had, but that doesn't mean we're out of the loop. In fact, you bring something AI never can: real-world wisdom and emotional guidance.

Here's how to stay involved and support your child's learning journey — even when the tech feels unfamiliar.

Get curious — not intimidated

AI in schools isn't about robots taking over. It's often small, behind-the-scenes tools helping with things like writing feedback, maths practice, or managing classroom tasks. You don't need to know every detail — just start by asking:

- Is AI used in your classroom? What for?
- Do you like the tools your school uses? Why or why not?

This shows your child you're paying attention — and it creates space for conversations, not lectures.

Ask your child's teacher

It's perfectly okay to ask your child's school how AI is being used. Some good starter questions:

- What kind of AI tools are part of learning here?
- How do you make sure students still think critically, even with tech involved?
- Are there safeguards around privacy or bias?

You don't have to challenge — just asking shows you care, and it helps you stay informed.

Encourage thinking, not just "getting the answer"

Tools like ChatGPT or automated feedback can be great — but only if they're used as a springboard, not a shortcut. Help your child slow down and reflect:

- What did the AI suggest?
- Do you agree with it?
- Could there be another way to solve that problem or explain that idea?

This builds the habit of critical thinking — a skill that will matter far more than memorising facts.

Talk about when AI helps — and when it doesn't

Make it normal to explore the strengths and limits of technology. For example:

- AI can help explain a maths problem — but it can't always tell why that answer matters.
- AI might summarise a story — but it can't feel what a character feels, or notice the deeper themes.
- AI might correct grammar — but it doesn't know your child's voice or creativity.

These are great moments to remind kids that they bring something valuable to every task: their ideas, their questions, and their judgement.

Set simple guidelines together

Rather than banning AI tools or handing them out freely, work with your child to set shared expectations. You might agree that:

- AI can be used to brainstorm ideas, but not to write full assignments.
- It's okay to use AI to check understanding, but not to copy answers.
- Screens have a time limit, and breaks are encouraged — especially when learning feels overwhelming.

Kids are more likely to respect rules they helped shape — especially when they understand the why behind them.

Model what good digital habits look like

Children learn more from what we do than what we say. If you're curious, thoughtful, and open about learning new things, they will be too.

- Share when you don't know something — and how you go about finding out.
- Be honest about when a tech shortcut seems tempting — and how you decide what's fair or responsible.
- Show them how to pause, question, and reflect — even when the answers come fast.

Just because it sounds confident, doesn't mean it's right.

A good reminder for every child using AI — and for all of us, really.

Questions for thought and conversation

As AI becomes part of your child's classroom experience, it's worth taking a moment to reflect — not with pressure to "get it right," but with curiosity and care. These prompts are here to spark conversation, either with your child or within yourself.

Being open, asking questions, and staying engaged is enough to make a big difference.

Try asking yourself:

- What tools are my child's school using — and do I understand what they do?
- Is my child learning with AI, or being taught by AI? How do I feel about that?
- How can I help them build confidence in their own thinking — not just rely on the answers AI gives?
- Am I creating space at home for curiosity, creativity, and digital downtime?
- What skills do I hope my child will carry into adulthood — and how are they developing those at school today?

These aren't one-time conversations. They're small check-ins that help your child feel supported, safe, and confident in a learning world that's changing fast.

Tools are changing — but kids still need us

AI is becoming part of everyday school life — from lesson planning and homework help to personalised learning support. But even as these tools grow more capable, they don't replace the human relationships that make learning meaningful.

What continues to matter most is the role we play in helping kids reflect, ask questions, and find their own way through the noise. When we stay connected to how learning is changing — and what our kids are experiencing — we're better placed to support them with care, confidence, and curiosity.

Because education isn't just about keeping up with new tools. It's about helping young people grow into thoughtful, resilient learners — ready for a future that's still being written.

In the next chapter, we'll explore what that looks like beyond the classroom — in our homes, relationships, and everyday decisions as parents raising kids in an AI-shaped world.

Chapter Four: Parenting in an AI World.

Parenting has always been about preparing kids for their future — not re-creating ours. But that's easier said than done when the world is changing this quickly, especially with the quiet rise of artificial intelligence in our everyday lives.

These days, AI isn't some distant, high-tech concept. It's here — in the videos our kids are recommended, the routes our phones suggest, the homework apps they tap on, and even the way we manage family calendars or set screen time limits. Sometimes it's helpful. Sometimes it's confusing. And often, it's invisible — shaping routines and decisions without us fully noticing.

That's exactly why it matters.

Because home is where children first build their relationship with technology. It's where they watch how we respond to tools that talk back, give advice, or suggest what to do next. They pick up on our comfort — or discomfort. Our curiosity — or hesitation. And most of all, our sense of when something is just convenient, and when it's actually worth paying attention to.

You don't need to be a tech expert. You don't need all the answers. But in a world shaped by AI, being a curious, present, and intentional parent makes a big difference. It helps kids grow up not just using technology — but thinking about it.

This chapter explores how AI is showing up in family life, the kinds of questions it raises, and how we can respond — not with fear, but with thoughtful conversations, shared learning, and a focus on what really matters.

Where AI shows up in everyday family life

You might not have a robot butler or a hologram teacher in your lounge room — but if you've ever asked Siri a question, followed Google Maps' fastest route, or let your child play with a learning app that "adapts" to their level, you've already welcomed AI into your home.

It's not always obvious. In fact, that's one of the most important things to understand: AI often works quietly in the background — making suggestions, tailoring recommendations, or answering questions — without ever announcing itself.

And for kids growing up now? That's normal.

They don't think of AI as something separate or special. It's just part of how their world works. From YouTube videos and streaming platforms to digital assistants and smart toys, they're surrounded by tools that respond, predict, and personalise.

It's important that we pay attention — not because AI is inherently good or bad, but because it's shaping how our kids learn, relax, communicate, and make sense of the world, often without much explanation.

Everyday AI in action

Here are a few familiar places where it shows up in family life:

Voice assistants (like Siri, Alexa, and Google Assistant)

"Hey Siri, how do you spell tyrannosaurus?"

"Okay Google, play Baby Shark!"

These voice tools use AI to understand language, interpret requests, and respond in real-time. For many kids, it's their first experience talking to technology — and expecting it to answer.

Why it matters:

When machines respond like people, it can blur the lines between what's smart and what's human. That's why it's helpful to talk with kids about things like tone, politeness, and the idea that not all answers are equally trustworthy — even when they sound confident.

Streaming services (like Netflix, YouTube, Spotify)

Ever wondered why YouTube seems to "know" what your child wants to watch next? AI recommendation engines track what's been viewed, liked, paused, or skipped — and serve up more of the same.

Why it matters:

This can create "filter bubbles" — where kids mostly see more of what they already like, but miss out on different viewpoints or fresh ideas. It's a great chance to talk about variety, trying new things, and pressing pause before letting autoplay take over.

Learning apps and games (like Reading Eggs, Prodigy, or Duolingo)

Some apps your child uses for maths, reading, or language now have AI built in. That means they can "adjust" as your child learns — offering easier or harder questions, hints, or feedback based on how they're going.

Why it matters:

This kind of learning can be helpful — especially if your child needs extra support or is ready to move ahead. But it's still important to check in. Some apps just focus on right answers, not real understanding. Kids still need conversations, encouragement, and guidance to make the most of what they're learning.

Online shopping and ads

From Coles remembering your grocery list to a toy ad popping up again and again, AI is behind the scenes, guessing what you might want to buy — and showing it to you (and your kids).

Why it matters:

For kids, ads can feel less like "selling" and more like friendly suggestions. Helping them notice what's being promoted — and why — builds early awareness. It's a simple way to start teaching them how to be thoughtful, not just reactive, consumers.

Safety and monitoring tools

Things like screen time limits, content filters, sleep trackers, or apps that show where your child is — many of these now use AI to notice patterns and make suggestions (like flagging late-night scrolling or unusual activity).

Why it matters:

These tools can help keep kids safe online. But they also bring up big questions — like how much is too much monitoring? As kids grow, involving them in decisions about tech helps build trust, responsibility, and a sense of ownership over their choices.

Homework help and chatbots

Tools like ChatGPT or AI-powered search engines are popping up more and more in how older kids do schoolwork. They might use them to get ideas, fix grammar, or even solve tricky maths problems.

Why it matters:

These tools can be helpful — but they can also become a shortcut. If kids rely on AI before giving something a real go, they miss out on the thinking part. Having open chats about when it's okay to use AI (and when it's not) helps build honesty and decision-making skills.

How AI shapes sibling interactions

AI doesn't just affect how kids interact with screens — it can also shape how they interact with each other. In families with more than one child, shared tech can quietly become a source of confusion, frustration, or even rivalry.

Think about it:

- A voice assistant that always answers the loudest voice.
- A streaming app that suggests shows based only on the older sibling's preferences.
- A homework tool that gives one child glowing feedback — and the other a string of red corrections.

None of these are dramatic on their own. But over time, small patterns like these can influence how siblings feel about each other — and themselves.

Why it matters:

- Subtle comparisons can build up. Kids may start to feel like AI tools "favour" one sibling over the other, especially if one gets more praise or seems to get better results.
- One child's voice can take over. The algorithms behind many tools adapt to whoever interacts with them most — often giving quieter or younger kids less influence over what's shown or suggested.
- Fairness feels harder to manage. If the tech seems to be working better for one child, it can make sharing — or even turn-taking — more emotional than expected.

So what can parents do?

Start by simply noticing how tech is shaping your family's shared spaces and experiences. Even small things — like whose profile is logged in, or who the algorithm seems to "like" more — can make a difference.

Try asking gentle, curious questions like:

- "Whose account is this logged into?"
- "Do those suggestions match what you actually like?"
- "How can we make sure everyone gets a fair go with this?"

It's just like helping kids share toys or take turns with the TV remote — only now, we're guiding them to share digital spaces thoughtfully too. Because even in a world full of personalised tech, what really shapes kids the most isn't the screen — it's their relationships.

Resource spotlight: Family Toolkit on AI

Day of AI Australia – Families Toolkit

A practical guide to help parents and carers explore AI with their children. Includes age-appropriate conversation starters, hands-on activities, and relatable examples to spark curiosity, build understanding, and encourage thoughtful discussion at home.

Access at:

dayofaiaustralia.com → Navigate to "Families Toolkit"

www.dayofaiaustralia.com/families-toolkit

It's not always obvious — but it's already here

One of the most surprising things about AI is how quietly it works in the background. Your child might not realise that a game is adjusting its difficulty, or that the next YouTube video, ad, or app recommendation is being chosen by an algorithm. To them, it just feels natural — like the app "gets" them.

When AI tools work invisibly, they can shape how kids see the world without them even realising it — what they watch, what they believe is "true," and how they solve problems. These aren't just learning tools or entertainment platforms — they're quietly shaping habits, values, and ways of thinking.

That's why it helps to bring some of that invisible influence out into the open.

We don't need to panic or hover. Just staying aware and making space for regular conversations — even casual ones — can make a real difference. A simple, "Why do you think it showed you that?" can spark awareness and give kids the tools to think for themselves.

Because once kids see how AI works, they're far more likely to use it wisely — instead of letting it quietly shape them.

Great questions to explore together:

- "Why do you think YouTube recommended that video?"
- "Do you think the answer Alexa gave was right?"
- "How do you know that information is true?"
- "When is it okay to use a chatbot, and when should you figure it out yourself?"

Opening the door to these conversations helps your child develop something even more valuable than digital skills: the ability to think critically, stay curious, and trust their own judgement in a world full of quick answers.

The big questions for parents

As AI becomes more woven into everyday family life, many parents find themselves wondering: What does this mean for my child? Am I keeping up? How do I guide them when I didn't grow up with this myself?

These are big questions — and you're not alone in asking them. You don't need to have all the answers, but having the right questions can make all the difference.

"Is this helping my child learn — or just helping them finish?"

AI tools can feel like magic. They speed things up, tidy grammar, solve equations, and summarise text in seconds. But learning isn't about fast answers — it's about understanding.

What really helps your child grow is:

- Struggling a little and figuring things out
- Making meaningful connections
- Asking "why?" and "how?"
- Learning how to learn — not just what to say

We don't want AI to become a shortcut that skips over the messy, meaningful parts of learning.

What to watch for:

- Is your child still able to explain their thinking in their own words?
- Are they turning to the AI tool before trying something themselves?
- Are they building confidence — or relying on the tool because they feel unsure?

It's fine to use AI — the key is to stay curious about how and when they're using it.

"Who's making decisions about the tools my child uses — and how?"

AI tools often come bundled into learning platforms, educational apps, or school-wide systems. Your child might be using them every day without anyone ever calling them "AI." That makes it harder to know what's going on — and even harder to ask the right questions.

As a parent, it's okay to ask:

- What tools is my child using — and what do they do?
- How were they picked?
- Have teachers had support in using them well?

- Is any of my child's data being collected — and how is it protected?

Many schools are still figuring this out too. Even if the answers aren't clear right away, asking these questions keeps the conversation open — and encourages schools to stay transparent.

"Is this fair, safe, and inclusive for every child?"

AI systems are built on data — and that data isn't always neutral. It reflects the world we live in, including its biases and blind spots.

This can show up in ways that affect real kids:

- A reading app might assume kids speak a certain dialect of English.
- An "engagement" tracker might penalise neurodivergent children who don't interact in typical ways.
- A recommendation algorithm might only surface content popular in dominant cultural groups.

When these tools influence decisions — about ability, behaviour, or progress — it can lead to unfair outcomes.

What you can do:

- Talk to your child about how tech works and who it works best for
- Raise questions with schools about inclusion and accessibility
- Let your child know it's okay to speak up if a tool feels unhelpful or confusing

Fairness in AI isn't just a tech issue — it's a values issue. And every child deserves to feel safe and understood in their learning journey.

"How do I set boundaries when AI feels invisible?"

Screen time is easy to spot. But AI? It's working behind the scenes in recommending shows, answering questions, helping with homework, tailoring games.

That invisibility makes it harder to manage.

Try setting gentle boundaries that build awareness:

"Try it yourself first" before using a chatbot

- Talk about how a tool was used after an assignment
- Pick tech-free times or spaces (like meals or bedrooms)
- Remind them: just because something's easier doesn't mean it's better

AI is designed to be helpful — but part of parenting in this new landscape is helping kids know when to accept help, and when to lean into their own effort and thinking.

"Am I supposed to be an expert in this?"

Not at all.

What matters most isn't knowing the technical stuff — it's being present, asking good questions, and helping your child think things through. You bring something no algorithm ever could: values, experience, and heart.

Simple questions like "How did that tool help?" or "Do you agree with that answer?" create powerful learning moments — for both of you.

Because good parenting in the AI age isn't about keeping up with the machines. It's about helping our kids stay connected to what matters most: their values, their voice, and their ability to think for themselves.

Building digital wisdom at home

Kids today are growing up with technology woven into nearly every part of their lives — and they often navigate it with impressive speed. But knowing how to use a tool isn't the same as knowing how to use it wisely.

That's where your guidance comes in.

You don't need to stay on top of every app or understand how every algorithm works. What really makes a difference is helping your child notice how tech affects them — their thinking, their focus, their choices — and giving them the confidence to pause, reflect, and steer their own path.

Digital wisdom means:

- Recognising when a tool is helpful — and when it might be doing the thinking for you.
- Being aware of what's shaping your feed, your mood, and your decisions.
- Knowing that just because something is fast or easy doesn't mean it's right.

What do we mean by digital wisdom?

Digital wisdom goes beyond knowing how to use technology — it's about helping our kids grow into thoughtful, aware, and grounded users of the digital world.

At its heart, digital wisdom is the ability to pause and think before reacting. It's being able to tell the difference between what's helpful and what's just convenient. It's noticing when a tool is making life easier — and when it might be doing the thinking for us.

It's understanding that:

- Not every chatbot response is correct.
- Not every recommendation is neutral.
- Not everything that's popular is worth trusting.

And maybe most importantly, it's knowing that technology doesn't come with values — we bring those. It's our compassion, patience, fairness, and judgment that help make tech use healthy and meaningful.

This kind of wisdom doesn't show up all at once. It's not something that can be downloaded like an app. It grows gradually, through daily life — through questions asked at the dinner table, the way we talk about a news story, or how we explain why a certain online trend doesn't sit right with us.

From rules to conversations

Setting boundaries around tech — like screen time limits or keeping devices out of bedrooms — is an important foundation. But when it comes to AI, it's no longer just about when or how long our kids are using devices. It's about how they're thinking about the tools they use.

That means shifting from rules alone to real conversations.

Instead of just saying:

"Don't use that app."

Try opening up a dialogue with questions like:

- "What do you think this app is doing with your data?"
- "How do you decide whether to trust what it's telling you?"
- "What would you do if the answer didn't feel right?"

These conversations don't need to be formal. You don't need a script or a special sit-down chat. Honestly, some of the best moments to talk are when you're doing something else — making dinner, driving to sport, walking the dog.

A quick, "What do you think?" or "Did that seem right to you?" can open up a meaningful moment.

Each time we help our kids slow down and reflect, we're doing something powerful. We're showing them that technology isn't just something to click through — it's something to notice, question, and use with care.

Because in a world where AI is working quietly behind the scenes — shaping what they see, learn, and enjoy — those everyday chats are what help them grow into thoughtful, confident users. Not just followers of tech, but guides in their own right.

Start early — but keep it age-appropriate

For younger kids (like preschoolers and early primary):

If you have a younger child, like a five-year-old, most of their interactions with AI won't feel like technology at all. It'll feel like magic. A voice assistant that answers their questions. A YouTube playlist that seems to know what they want to watch. A maths game that gets easier (or more fun) the more they play.

They're not thinking about algorithms or data. They just see something that works — and responds.

That's why this is a great time to start laying the groundwork. Not with big explanations or warnings, but with gentle curiosity and small observations that build awareness over time.

Try:

- "That's a robot voice — it doesn't really know you, it's just guessing what you might want."
- "Why do you think it showed that video next? Do you like that one too?"
- "Let's ask it the same question again — does it give the same answer?"

These kinds of questions help your child learn that not everything they see is random or neutral — and that it's okay (and even good!) to be curious about how things work.

You're not trying to make them sceptical or anxious. You're giving them the early tools of digital thinking: noticing patterns, asking why, and knowing that their ideas and instincts matter. These early lessons will grow with them — and make it easier to have deeper conversations later on.

For older kids and teens:

As kids grow, so does their curiosity — and their independence. By the time they're in upper primary or high school, many are already experimenting with AI in ways that feel helpful, exciting, or even a little secretive. They might use it to summarise a tricky homework passage, plan a speech, or get help with writing. Some might explore AI for music, art, or even chatting. And yes — some won't always tell you they're using it.

That's completely normal.

This is a stage where conversations shift from control to guidance. Instead of trying to monitor everything, it becomes more about helping them think critically and make wise choices on their own.

You might ask:

- "Would you trust AI to give someone a grade or a score?"
- "How do you decide if what it tells you is actually right?"
- "If a chatbot helped you write something, do you still feel like it's yours?"
- "Have you ever seen it get something really wrong?"

Even if your teen shrugs or rolls their eyes, these questions matter. They plant seeds. They say: "I trust you to think about this." And over time, that trust helps build the kind of inner compass that's much more powerful than any parental control setting.

Because our goal isn't to block every misstep — it's to help our kids learn how to navigate a world where AI is part of the landscape, and values like honesty, responsibility, and fairness still count.

Teach the difference between smart tools and smart thinking

AI tools can seem pretty amazing. They can help write essays, solve maths problems, suggest what to watch, or even chat like a real person. But here's the thing: they don't understand the way we do. They're not thinking — they're just really good at guessing what might come next, based on patterns in data.

That's why it's important to help kids see the difference between using a smart tool — and thinking smartly.

You might say:

- "This tool is great for speeding things up — but does the answer actually make sense?"
- "What do you think before we ask the chatbot?"
- "That was a clever suggestion — but do you agree with it?"

Here's a simple way to break it down:

- AI is good at: helping with tasks, sorting information, making quick suggestions.
- Humans are good at: judgment, creativity, empathy, fairness — and knowing when something just feels off.

Even when a tool sounds confident, it doesn't mean it's right. Helping your child notice the difference builds the kind of awareness they'll use over and over again — in school, online, and in life.

Model curiosity, not perfection

Your child doesn't need you to have all the answers — just to show that you're thinking it through too.

Try things like:

- "Hmm, I'm not exactly sure how that works — want to figure it out together?"
- "That answer came up really quickly... but does it feel right to you?"
- "Interesting choice from the AI — what do you think made it suggest that?"

These kinds of comments may seem small, but they're powerful. They show your child that being thoughtful matters more than being right. That it's normal (and smart!) to pause, check, and wonder.

Over time, you're helping them build habits of reflection, critical thinking, and self-trust — not just around technology, but in how they approach the world.

Make values part of the conversation

AI tools might seem neutral — just smart machines doing their thing. But behind the scenes, they're shaped by people: built by humans, trained on human data, and often designed to meet business goals. That means they can reflect the values and assumptions of their creators — and those aren't always fair or inclusive.

This is why values matter. Not everything AI suggests is worth copying. And just because a tool can do something doesn't mean it should.

You can help your child think it through by asking:

- "Does this seem fair?"
- "Is this helping you be kind and honest?"
- "Would you be okay if someone else used it this way?"

Conversations like these help kids learn that tech use isn't just about what's possible — it's about making thoughtful, values-based choices.

Raising digital citizens, not just digital users

It's one thing to know how to use technology — to tap the app, follow the prompts, or find the answer fast. But it's another thing entirely to understand what's happening underneath: how those tools are designed, what they're encouraging, and who they're serving.

That's the difference between being a digital user and a digital citizen.

Digital citizens are aware. They ask questions. They think about fairness, safety, and impact — not just for themselves, but for others too. They don't just use tech to get things done. They use it with purpose and care.

And just like any kind of citizenship, these habits start at home — with small conversations, shared questions, and the example we set every day.

From consumers to citizens

Most of the tech our kids use — from games and streaming platforms to social media and shopping apps — isn't just designed to work. It's designed to keep them coming back. Features like auto-play, streaks, endless scroll, and push notifications are all built with one goal in mind: holding attention.

And while that can make life more convenient, it can also quietly shape habits, shorten attention spans, and make it harder to switch off.

That's why we need to help kids go from just using tech to thinking about how it works — and why it works the way it does.

Try asking:

- "Why do you think that video played right after the last one?"
- "Do you think this game wants you to stop — or keep going?"
- "What would you change to make it feel more fair or balanced?"

These kinds of questions help kids take a step back — and start seeing themselves not just as users, but as active participants in the digital world.

Simple ideas to start with

You don't need a big plan or fancy tech rules to help your child build digital wisdom. Just start where you are — and let everyday moments do the work.

Try this:

- During homework: Ask how they tackled the task — and if any apps or tools gave them a hand.
- After gaming: Chat about how the game changed as they played. "Did it get harder? Was it reading your moves?"

- While watching videos: Wonder aloud, “Why do you think it showed that one next?”
- At the dinner table: Share a story about a time you learned something the long way — and how it helped you remember it better.

These little conversations may seem simple, but they help build the habit of thinking before clicking — and that’s a skill that lasts.

Bringing it all together

Let’s be honest — keeping up with AI can feel like a full-time job. The tech changes fast, and it’s easy to feel like we’re always one step behind. But the good news? You don’t need to be a tech expert to raise kids who can thrive in this new world.

What really matters is being present. Asking questions. Talking about what’s happening — even when you’re not totally sure yourself. That’s how digital wisdom grows: not through perfect answers, but through open, curious conversations.

As we’ve seen throughout this chapter, AI isn’t some far-off thing. It’s already part of how our kids learn, play, explore, and make sense of the world around them. That’s why our role as parents is so important — not just to keep them safe, but to help them grow into thoughtful, confident people who can navigate tech with care and make choices they feel good about.

What can that look like in everyday life?

- It might be taking a few minutes to ask, “Hey, how does that app actually work?” — not to test your child, but to show you care about what they’re learning.
- It might sound like, “Hmm, I’m not sure either. Want to figure it out together?” — and showing them that learning together is just as valuable as knowing the answer.

- It means weaving your family's values into everyday tech use — not as a one-off rule, but as an ongoing chat about fairness, honesty, and what feels right.

These small moments build trust and help your child see that tech isn't just something they use — it's something they can think about, question, and use wisely.

Using AI as a parenting ally (without over-relying on it)

Just as AI shows up in kids' lives, it's also finding its way into our parenting routines—whether it's helping plan meals, keeping track of bedtime routines, or giving a hand with task lists. These tools can free up mental space, letting us be more present with our kids when it counts. But there's also a subtle tug toward over-optimizing everything—and that's when tools designed to help can start feeling like pressure instead.

A simple, steady touch for balance:

Before letting a tool take over, ask yourself: Is this giving me back time and calm—or just tacking on a new expectation?

AI can be helpful when:

- It creates breathing room—like summarising a complex email so you can focus on connecting with your child.
- It supports routine—like helping with bedtime rhythms or homework reminders.
- It explains things in a way that actually connects—translating a school note or giving a gentle how-to.

But it's also okay to say "not today" to certain features. Just because an app can track every Just because an app can track every grunt, giggle, or sigh doesn't mean it should. Your radar—your gut feeling about what's too much—is powerful. Trust that.

As a parent you are in the position to have what your child needs most:

- A steady presence in a world that's always changing

- A willingness to learn and figure things out together
- And the kind of example that shows what thoughtful tech use really looks like

There will always be new apps, tools, and challenges — that's just part of growing up today. But when we focus on what really matters — strong relationships, thoughtful reflection, and clear values — we give our kids something solid to stand on.

So as you move forward, don't feel like you need to have it all figured out. Stay curious. Stay connected. And above all, keep showing up — because your presence and perspective are the most powerful tools your child will ever have.

Chapter Five: Tech, Mental Health & Wellbeing in an AI World

Emotional resilience, social pressures, and healthy tech use

We spend a lot of time thinking about how AI will change school or work — but what about how it's quietly shaping our kids emotionally and socially? It's not always obvious. It's in the little things: the pressure to stay connected, the algorithm that decides what video plays next, or the way a chatbot responds to a late-night worry. These moments may seem small, but over time, they help shape how kids see the world — and how they see themselves.

AI now plays a bigger role in children's mental wellbeing than many of us realise. Social feeds adjust to keep them scrolling. Games track their progress and praise them just right. Even "wellness" tools can offer advice based on their tone or typing speed. These experiences can feel helpful — even comforting — but they can also subtly influence confidence, self-image, and emotional habits.

And all this is happening at a time when kids are already growing up with more pressure than ever. From climate change to cost-of-living stress, to the non-stop noise of digital life, today's young people are navigating a world full of uncertainty and complexity. That's why emotional resilience is so important — the ability to understand their feelings, manage stress, and stay centred when things feel overwhelming.

Here in Australia, the federal government is now moving to ban social media for kids under 16. It's a strong response to growing concerns — not just about what children see online, but how algorithm-driven platforms can shape their emotions, behaviour, and even identity. These aren't neutral spaces. They're designed to hook attention and keep users coming back — especially younger ones.

That's why emotional resilience matters more than ever — the ability to understand their feelings, manage stress, and stay centred when things feel overwhelming. In a world where tech is designed to influence, helping kids stay grounded is just as important as helping them stay safe.

This chapter explores where tech and emotions meet — and what we can do at home to support mentally strong, emotionally aware kids in an AI-shaped world.

Where AI touches kids' mental health

AI might sound like something out of a sci-fi movie — something that lives in labs or big tech offices. But for our kids, it's already part of daily life. It's woven into the apps they use, the games they play, and the content they scroll through. And while it might not seem obvious, it can have a big impact on how they feel about themselves.

We're not just talking about screen time here. We're talking about how AI shapes what kids see, how they compare themselves to others, and what messages they pick up about things like success, beauty, or popularity — often without even realising it.

How social media algorithms shape what kids see

Social media platforms use AI to decide what shows up in your child's feed — and when. These systems are designed to keep people scrolling, not to protect their mental health.

So what does that mean?

- Posts that stir up strong feelings — like jealousy, anger, or FOMO — often rise to the top.

- Filters and polished photos can make it seem like everyone else's life is more exciting or perfect.
- Likes, comments, and views can feel like a scoreboard for self-worth.

Even kids who are usually confident can start feeling "not good enough" after too much scrolling. And for kids who are already feeling anxious or unsure, it can quietly make things worse.

Things to Keep an Eye On:

- Is your child more self-conscious about how they look?
- Do they talk a lot about likes, followers, or going viral?
- Are they in a low mood after using certain apps?

If any of that sounds familiar, it doesn't mean you need to panic. But it's a good reason to pause, check in, and have a gentle conversation.

Gaming & AI: more than just play

These days, a lot of the games our kids love are powered by AI behind the scenes. They adjust as your child plays — making things easier or harder, suggesting new tools or upgrades, and gently steering them toward certain choices.

That can make games more fun and engaging — but there's a flip side too.

Some games are designed to keep kids playing for longer. They use things like streaks, surprise rewards, or "just one more level" tricks to keep the momentum going. And while that can look like fun, it's actually a form of emotional design — where the game is tapping into your child's feelings to keep them coming back.

Here's what to watch for:

- Games that feel hard to stop, even when your child is tired or frustrated
- Big mood swings after losing or winning
- Pressure to buy extras to keep up or feel successful

It's not about banning games — consider how how they're built, and help your child recognise how they're feeling. Some games teach great skills like patience, problem-solving, or creativity. But others can spark stress, comparison, or even low self-esteem if the design is too intense.

Keep the conversation open: "How did that game make you feel today?" or "What do you like about this one?" goes a long way.

Chatbots, companions, and AI as a "friend"

Some kids and teens are starting to chat with AI bots — like Replika, Character.ai, or even Snapchat's My AI. These bots are designed to feel friendly, helpful, and sometimes even a bit like a "virtual friend."

And for a young person who's feeling lonely, stressed, or just wants to talk without pressure, that can be really appealing.

But there are a few important things to keep in mind:

- These bots might sound kind or supportive, but they don't actually understand emotions — they're just mimicking patterns in data.
- The conversations might feel meaningful, but they're not real relationships. That can sometimes blur the lines between connection and confusion.
- Some bots, especially the less-moderated ones, can share content that's inappropriate — even if they seem harmless at first.

If your child is using these tools, try not to panic. Instead, be curious. You might say:

- "What do you like about chatting with it?"
- "Has it ever said something weird or made you feel uncomfortable?"
- "Do you think it really 'gets' you, or does it just seem like it does?"

Don't ban or shame — but rather help kids think about what's real, what's helpful, and what healthy connection actually looks like.

Personalised feeds & mood loops

Apps like YouTube, TikTok, Spotify, and even Google are built to "learn" what we like — and then show us more of it. That might mean more funny videos, more chill music, or more how-to guides on something your child's curious about.

But here's the catch: these tools aren't always tuned into how a child is feeling — they're tuned into what they click.

So if your child watches one sad or intense video, the app might think, "Oh, you want more of that!" And suddenly their feed is full of moody songs, scary news, or anxious content.

It can create an emotional loop, without them even noticing:

- They feel a bit down → click on something emotional → get fed more of the same → start feeling worse.

Or they might go deep into a topic — like extreme diets or conspiracy videos — without seeing other viewpoints to balance it out.

That's the thing about AI: it's not trying to be helpful or harmful. It just keeps serving what seems to "work" — whether it's good for your child's wellbeing or not.

That's why it helps to check in sometimes with gentle questions like:

- "What kind of stuff has your feed been showing lately?"
- "Does it lift your mood, or make you feel more stressed?"
- "Want to watch or listen to something different together?"

Why This Matters

Supporting our kids' mental health isn't just about blocking the "bad stuff." It's about making sure they feel understood, confident, and connected — especially in a digital world where so much is quietly shaped by algorithms.

When kids understand how these tools work, they're less likely to be swept along by them. They start to notice what makes them feel good... and what doesn't. And that awareness? It's powerful.

Because once they can see how tech affects their thoughts and moods, they're better equipped to make choices that actually support their wellbeing — not just keep them scrolling.

The pressures of being "always on"

Remember when the school day ended at 3pm and that was it? You could go home, unwind, and leave all the social stuff behind until tomorrow.

That's not how it works for kids today.

Their social world follows them home — buzzing, pinging, and updating non-stop. And a lot of that pressure comes from how the platforms they use are built. Algorithms are constantly working in the background, nudging them to stay online: "Watch this next," "Don't miss this update," "Why haven't you replied?"

It may not be obvious or in-your-face, but it never really stops. And for kids, that can be exhausting.

The social pressure to be "available"

In group chats, games, or apps like Snapchat and Instagram, there's often this constant expectation:: if you're online, you should reply.

- They might worry about upsetting a friend if they don't respond fast enough.
- "Seen" notifications create silent accountability — They know I read it, now I need to say something.
- Some platforms even show when someone was "last active," adding to the sense of social expectation.

From the outside, it might look like they're glued to their phone or being dramatic. But often, it's really just social anxiety. They're doing their best to keep up with friendships — on platforms designed to never take a break.

AI makes the internet feel endless

Ever notice how one video turns into ten? Or how a game that was meant to be a quick break suddenly eats up an hour? That's not by accident. AI tools are built to keep kids (and all of us) clicking, scrolling, and staying engaged.

- There's no "you're done" — just the next video, the next post, the next level.
- The more your child watches or plays, the more the algorithm learns what hooks them — and feeds them more of it.
- Even games adjust in real time to keep things exciting — not too easy, not too hard, just enough to keep them saying, "one more go."

It's easy for kids to lose track of time or feel drained without realising why. Their brains are working overtime, constantly responding to stimulation — and that can take a toll, even when everything seems "fun."

The "fear of missing out" is real — and amplified

AI doesn't just show your child what they like — it shows them what others are doing, watching, or celebrating. That can make life feel like one big competition they didn't sign up for.

- "Everyone's playing this game — if I don't join, I'll be out of the loop."
- "That post got hundreds of likes. What's wrong with mine?"
- "If I miss the chat tonight, I'll be left out tomorrow."

This can lead to a constant sense of urgency — like they have to be online or risk missing something important. It's not just about fitting in. It's about how these platforms make "falling behind" feel like a real threat, even when it's not.

Notifications = micro-stress

We often talk about screen time — but there's something just as important that flies under the radar: the constant interruptions that come with it. Every ping, buzz, banner, or alert might seem small in the moment, but over time, they chip away at our kids' ability to focus.

This leads to attention fragmentation — when the brain is constantly switching between apps, messages, and tasks, instead of staying with one thing. That kind of scattered focus makes it harder to finish homework, stay present, or even unwind properly.

Here's how it shows up:

- Notifications pull kids out of what they're doing — even during quiet time or study.
- Their brains start to expect interruptions, making deep focus harder to achieve.
- Some even experience "phantom buzzes" — feeling like their phone vibrated when it didn't.

On their own, these things might seem harmless. But together, they create a steady stream of micro-stress — leaving kids feeling wired, jumpy, or just unable to fully relax. It's not just about how much time they spend on screens — it's about how often they're being pulled away from whatever they're trying to do.

Bedtime battles and digital wind-down

If your child resists bedtime or seems wired even when they're clearly tired, you're not alone. We often think of this as a "rules" issue — but AI-driven content plays a bigger role than we realise.

Many videos, games, and apps are designed to grab attention quickly — with fast cuts, bright colours, emotional hooks, or just plain silliness.

That might be fun earlier in the day, but it works against the brain's natural wind-down process at night.

Here's what's happening:

- Blue light from screens can delay the release of melatonin – the hormone that tells the brain it's time to sleep.
- Exciting or emotional content keeps the brain in "alert mode," even if your child says they're sleepy.
- And for older kids and teens, late-night scrolling can feel like their only quiet time – which makes it harder to let go.

That's why helping kids wind down with gentler, screen-free routines can really help. A warm bath, quiet music, a familiar bedtime story – these kinds of rituals signal to the body and brain that it's safe to slow down. It's not about being strict for the sake of it – it's about giving their minds the space they need to rest.

What you can do

You don't have to go to war with technology – but it helps to slow things down now and then. Just making a bit of space to talk and notice how it all feels can make a big difference.

Try asking your child:

- "Do you feel like your phone's helping you – or just keeping you busy?"
- "What helps you feel calm after spending time online?"
- "Is there one notification you'd be happy to turn off this week?"

You could also try setting up calm corners at home or create mini unplugged rituals before bed. It's not about being perfect – it's about building little moments of pause, awareness, and balance into everyday life.

Helping kids build emotional resilience

AI might be changing how our kids learn, connect, and play — but it's what's going on inside that really shapes who they become. Emotional resilience is the ability to bounce back, handle change, and stay steady when things get tough.

But in a world full of pings, scrolls, and always-on tech, those steady moments are getting harder to find. So much of what kids see online is instant, polished, and perfect — and that can make it easy to forget that real emotional strength grows slowly. It comes from trying, failing, talking things through, and learning how to sit with tricky feelings.

So what can we do, as parents, to support that kind of growth at home? Let's explore a few simple ways.

Notice what's getting in the way

AI and modern tech bring emotional challenges that most of us never faced growing up. For example:

Highlight reels: Kids often compare their everyday lives to the polished, curated content served up by algorithms — and that can quietly chip away at their confidence.

Quick fixes: AI tools are designed to keep things moving. They often distract rather than help kids sit with big feelings.

Outsourced comfort: Mood trackers and chatbots might feel helpful, but they can't replace a real conversation or a comforting hug.

What you can do

Name emotions out loud — not just the tech habits.

Try simple observations like:

- "You seemed really flat after that game — what do you think was going on there?"
- "That post looked like it hit a nerve. Want to talk about why?"

Show that hard feelings are normal.

- Let them know it's okay to feel left out, disappointed, or frustrated.
- "That looked like it really stung. You're allowed to feel upset."
- "I've had days like that too — it's tough."

Normalise imperfection and growth.

AI might feel seamless, but real life isn't. Kids need to see that it's okay to mess up — and recover.

> "I completely stuffed something up today — want to hear how I sorted it out?"

Create space for 'nothing'.

Unstructured time gives kids room to unwind and check in with themselves. It doesn't need to be productive — just quiet, unscheduled space to breathe.

With teens: keep the door open

Teens live in a world of AI-enhanced trends, digital comparison, and pressure to be constantly connected. They might not always want to open up, but they still need to feel seen and supported.

Lead with curiosity, not control.

- "That AI tool seems clever. Do you think it's helpful or a bit pressure-y?"
- "Do your friends use it too, or is it more your thing?"
- Respect their space, but stay gently present.
- Reinforce the message: their feelings matter, their questions are valid, and no one expects them to have everything figured out.

Healthy tech boundaries — without shame or fear

Setting limits around tech can feel like a tug-of-war: either you're constantly saying no, or you give in because it's just easier. Add AI into the mix — with its sneaky nudges and personalised suggestions — and it can feel like the screens are running the show.

But here's the good news: you don't need to ban devices or fight every night about screen time. Creating healthy boundaries is more about building habits that reflect your family's values — and helping your child understand why those habits matter.

Shift from control to connection

Strict rules might work in the moment, but they don't teach kids to tune into how tech actually affects them. Instead of just laying down the law, try talking about what's going on behind the screen — and how it makes them feel.

Ask questions like:

- "When do you find it hardest to switch off? What helps?"
- "How do you feel after a big scroll or gaming session?"
- "Do you think that app helps you chill out — or just winds you up?"

These kinds of chats invite reflection — without blame.

Let your values lead

Great tech habits start with knowing what matters most in your family.

Try asking yourself:

- What kind of mornings do we want to have?
- How do we want to connect at dinner or before bed?
- When do we all need a break from screens — just to be?

Then turn those values into small rituals:

- "We unplug an hour before bed so our brains can slow down."

- "We eat without phones so we can talk — even if it's about nothing much."
- "Sundays are our screen-light days — for nature, rest, or just some boredom."

When the rules make sense, they're easier to stick with (and easier to revisit as kids grow).

Don't rely on guilt to get compliance

It's tempting to use guilt ("You're always on that thing!"), but shame rarely leads to better decisions — just better hiding. Instead, try staying curious, even when you're concerned.

Swap this:

"You're addicted to that game!"

to

"What do you like about it? How does it feel when you stop?"

That tone of openness keeps communication alive and builds emotional intelligence.

Use smart tools — wisely

Ironically, some of the best tech boundaries are supported by tech itself. Tools like:

- Screen time trackers (Google Family Link, Apple Screen Time)
- Focus modes that limit distractions during homework
- App timers that nudge breaks during long sessions

But don't just turn them on — use them with your child.

Show them how to:

- Check their own screen time stats
- Set reminders to pause and stretch
- Use wind-down settings before bed

It's not about spying. It's about helping them notice their own patterns and feel more in control.

Talk about the AI inside the apps

AI is already baked into most apps your child uses — whether it's choosing videos, adjusting difficulty, or recommending what to do next.

So when you talk about boundaries, go beyond time. Get curious about how and why things are designed.

Ask:

- "What do you think this app wants you to do?"
- "Does it help you feel creative — or just distracted?"
- "Do you feel better or worse after using it?"

You're not critiquing — you're helping your child build digital instincts.

Your presence matters more than any rule

There's no perfect system. Boundaries will be bent, rules forgotten. But your presence — steady, caring, curious — is more important than any app or timer.

Be the person who asks:

- "What's that new AI art app? Want to show me what you've made?"
- "That chatbot sounds clever. Was it helpful, or did it miss the mark?"
- "Do you ever feel like you've had enough tech time for one day?"

These aren't "monitoring" moments — they're connection moments. And over time, those small moments are what make boundaries stick.

Signs your child might be struggling — and what to do

Even in tech-aware households, it's not always easy to tell when technology is affecting a child's mental health. Kids rarely say, "This app is making me anxious," or "I feel overwhelmed by what I see online." Instead, the signs often show up in quieter ways — in their sleep patterns, mood, confidence, or behaviour.

What to watch for

Even in homes where tech use is thoughtful and well-managed, it's tricky to notice when an app, a feed, or a game is starting to weigh more than it entertains. Kids don't usually say, "This app is stressing me out" — instead, those feelings show up quietly. And because kids are so adaptively tough, it's often easy to miss.

Here are some of the softer signs that might hint something's not quite right:

- Changes in mood or energy after screen time

  They go from lively to flat. Or suddenly irritable, quiet, tearful, or restless — especially after scrolling or gaming.

- Pulling away from things they used to enjoy

  Once-loved hobbies or hanging out with friends or family start feeling optional — or even overwhelming.

- Trouble sleeping

  Difficulty falling or staying asleep — especially when devices are part of a nighttime routine.

- Fixation on certain content

Whether it's about beauty, popularity, or knowing the "coolest" trends, obsession with these areas can start to feel out of proportion.

- Secretive or anxious screen use

Hiding what they're watching, trying to "catch up" fast, or seeming upset after a scrolling session can signal something's off.

- Constant self-comparison

Comparing their life, body, or skills to influencers, online friends, or curated content — and feeling short in comparison.

- Using screens to escape big feelings

Gaming, watching, or scrolling to avoid sadness, boredom, or anxiety.

A tough reality — and a powerful response

Australia has seen some heartbreaking stories in recent years. Young people have been seriously harmed by online experiences — everything from bullying and exploitation to anxiety driven by social media. These stories aren't just headlines; they have led to real change.

What's changing :

- The eSafety Commissioner now has stronger powers to demand the removal of harmful content — no matter where it originates — including violent, exploitative, or self-harm material.
- Social Media Platforms like Instagram, TikTok, Snapchat, and even YouTube must soon prevent users under 16 from having accounts, or face steep fines — up to around AUD $50 million.
- Australia is also setting up new industry codes aimed at protecting kids from age-inappropriate content related to self-harm, eating disorders, and more.

What this means for parents

1. Action is happening — not just talk.

Australia has moved from discussion into actual legislation and enforcement. That means tech companies are now being held to account — and safety is no longer optional.

2. Responsibility is shifting — to companies, not kids.

Parents aren't on their own here. The law places the onus on platforms to keep kids safe — legally and technically.

3. It's part of a broader push to protect kids' mental wellbeing.

The latest policy doesn't just lock kids out of screens — it gives parents more time and space to guide them thoughtfuly. The aim is to delay exposure and support growth and digital literacy.

4. The systems are playing catch-up — so families still matter.

Even as laws evolve, parents still need to translate protection into connection. Policies signal change. What really helps is what happens in the home—ongoing conversation, awareness, and care.

How to bring this into your parenting

Talk about it together. Let your kids know this isn't just a rule from the government — it's a sign that Australia is taking their mental health seriously.

- Let them feel included in the conversation. Ask:
- "What do you think about new laws stopping kids under 16 from being on social media?"
- "Do you feel like it gives you space to figure things out — or makes you feel excluded?"
- Use the change as an opportunity to guide, not block. If your child is disappointed, you could say, "I get why it feels frustrating — but it gives us time to focus on other things that build strength, not reliance."

This is a real moment — complex, imperfect, but full of potential. By staying curious, connected, and grounded in what matters most — relationships, reflection, resilience — you help your child grow, not just through this change, but because of it.

What you can do — with compassion

You don't need to have all the answers. But you do play a powerful role in noticing, listening, and supporting your child if they're struggling.

Try this:

- Open conversations without judgment. "Hey, I've noticed you seem a bit low after being online — want to talk about it?"
- Be specific and caring. "I saw that message you got — that would have upset me too. Do you want to chat, or just hang out for a bit?"
- Model your own boundaries. Let them see you taking breaks, talking about online pressure, or questioning what you see too.
- Remind them they're not alone. "Lots of people feel like this. It's okay to talk about it — we'll figure it out together."

If things feel bigger than you can handle alone, reach out. You can speak to your GP, school counsellor, or organisations like:

- Kids Helpline (1800 55 1800) — free, confidential support for kids and teens 24/7
- Headspace — youth mental health support across Australia
- The eSafety Commissioner — help with online safety, including cyberbullying and image-based abuse

Above all, remind your child that they are not the problem. If an app, game, or experience is affecting them, it's not about weakness — it's about understanding how powerful these technologies are, and learning how to navigate them with support.

When kids feel replaced by AI

Using AI tools—like having help with homework or getting quick feedback—can make learning easier. But it's a different thing entirely when kids start feeling like they're being measured against these tools—or even replaced by them.

At school, AI might grade an essay, flag plagiarism, or offer tutoring. At home, it may proofread stories or run math problems in seconds. While these tools are impressive, they can also unintentionally send messages like:

- "The AI is better at this than I am."
- "Why bother when the robot does it faster?"
- "Do I even matter anymore?"

For kids who are already unsure of themselves, these thoughts can sneak in quietly. A five-minute scroll past polished AI-written essays can spark questions like:

- "If AI can write an essay in seconds... why should I try?"
- "If I can't do it as fast or as well as the tool, what am I good at?"

Even high-achievers feel the pressure. They may worry they must always be more ahead, more efficient, more polished—often with AI's help. That can leave no room for mistakes, for experimenting, or simply being human.

What parents can do

Remind them of human value over speed or polish.

Try saying: "AI can write fast—but it can't bring your voice or your ideas." Or: "Mistakes aren't failures. They're how we grow."

Celebrate the beautifully flawed moments.

Whether it's a messy doodle, a half-finished draft, or an experiment gone sideways—laugh about it, wonder about it, ask what it was trying to teach rather than hiding or fixing it right away.

Reframe AI as a tool, not a standard.

You might say: "Think of AI like a calculator—not the person. It helps, but it doesn't replace your thinking." Or: "If we let the tool do the thinking, we don't get to do the learning."

Make failure and trial part of everyday conversation.

Share your own messy learning experiences: "I tried cooking that recipe—it turned into its own kind of science experiment!" These moments let kids see that experimenting and failing are just part of life, not a sign of weakness.

When we remind kids—and ourselves—that being human means being curious, messy, creative, and resilient, we help them hold onto what AI can never take: their voice, their value, and their potential.

Supporting kids through AI-influenced anxiety

Even if your child isn't talking about AI directly, it might still be shaping their sense of the world — and sometimes, their worries.

Whether it's a fear of being left behind, unease about how fast things are changing, or anxiety about a future that feels uncertain, AI can be an invisible source of stress.

And for some children, especially teens, these feelings may blend with broader mental health concerns like eco-anxiety, climate grief, or fear about jobs and security.

They might not say, "I'm worried about AI."

Instead, it might sound like:

- "What's the point of school if AI is doing everything?"
- "I'll never be as smart as that."
- "It's all moving too fast."

These are real, valid feelings — and it's okay if you don't have all the answers. What matters most is creating a space where kids can talk, be heard, and feel supported.

What helps

- Validate, don't dismiss.

"Yeah, it is a lot to take in. Want to talk more about what's on your mind?"

- Name what's human.

"The world's changing — but the things that really matter, like kindness and connection, don't go away."

- Keep the future open.

"We don't know exactly what jobs or tools will exist — but I believe in your ability to learn and adapt. That's what really counts."

- Don't avoid the topic.

Avoiding "big, scary" things can sometimes make them loom larger. Talking about AI — gently and age-appropriately — helps kids make sense of what they're hearing.

- Focus on action.

Helping kids see where they have agency — whether that's setting limits on tech, making art, joining discussions, or just unplugging for a while — can build resilience and confidence.

It's important to pause here and say: it's not all doom and gloom.

Yes, AI is reshaping the emotional landscape our kids are growing up in — sometimes in ways that raise serious concerns. But it's also true that not every child is overwhelmed or misled by it. Many are starting to ask great questions, spot the patterns, and think critically about how AI fits into their lives.

And that's encouraging.

Kids aren't all reacting the same way — and that's the point. Some may feel confused, anxious, or even defeated by AI's presence in school or social media. Others might be fascinated, empowered, or simply curious. Often, they feel a mix of all these things.

Questions for thoughts and reflection

As you think about your child's mental health and wellbeing in a world shaped by AI, these questions can help guide your parenting — not with judgment, but with curiosity and care:

- How does technology make my child feel — not just what does it help them do?
- When do I notice signs that they're overwhelmed, overstimulated, or anxious?
- Have we talked about how social media works — including the role of algorithms?
- What emotional habits are we modelling when it comes to stress, attention, or screen time?
- How can we make more space for rest, boredom, creativity, or outdoor time?
- Am I creating a safe space for my child to talk about fears — even ones I can't fix?
- What conversations do I want to keep open as my child grows into this new world?

These questions aren't about perfect answers. They're about being present, staying engaged, and reminding your child — and yourself — that while the tech may be smart, what matters most is the human connection at home.

Disclaimer:

I'm not a psychologist or mental health professional — I'm a parent doing my best to understand this rapidly changing world, just like you. The ideas in this section are based on current research, expert insights, and conversations with other parents. They're not a substitute for professional advice. If you're worried about your child's wellbeing, please reach out to your GP, a school counsellor, or a qualified mental health service for support.

Chapter Six:
Play, Creativity & Storytelling in the Age of AI

How AI is changing imagination, self-expression and childhood play

When you think back on childhood, what stands out?

Maybe it's those cozy storytelling moments, paint-splattered art sessions, backyard hunts for hidden treasures, or building entire worlds out of LEGOs. All of these moments spark something special: imagination.

But now, imagination is bumping into something new — artificial intelligence.

From apps that write poems or sketch stories, to talking toys and games that adapt as your child plays, AI is quietly finding its way into how kids create, play, and express themselves. Sometimes that opens up exciting new doors. Other times, it can feel like the creative spark is doing the imagining for them.

This chapter is about keeping that spark alive — making sure creativity stays firmly in our kids' hands (and heads).

Here's what we'll explore together:

- How AI is showing up — in toys, games, and creative apps
- What helps boost true imagination, and what might hold it back
- Why unstructured, messy play still matters more than ever
- Simple ways parents can foster creative confidence in the AI age

Because sure, AI can generate. But our kids are the ones who truly imagine — and that's something worth celebrating.

AI in games, toys & play

You might not always notice it, but AI is already part of how many kids play. It's in the smart speaker that answers their questions, the game that gets harder or easier depending on how they're doing, or the toy that remembers their name and favourite colour. These aren't just cool gadgets — they're shaping how kids explore, imagine, and interact with the world.

Where you'll see AI in everyday play

Here are some familiar examples:

- Voice Assistants (like Alexa, Siri or Google): Lots of kids love asking them questions, playing music, or hearing jokes. It can feel like magic — but it also teaches kids to expect quick, confident answers (even when those answers aren't always right).
- Smart Toys: These are dolls, robots or talking animals that respond to your child's voice, remember their likes, and sometimes hold short conversations. They can be fun and engaging — but they also raise questions about privacy and whether kids start relying too much on tech that always responds.
- Gaming Platforms: AI is what helps many games adjust to your child's skill level, suggest new challenges, or recommend what to play next. That keeps the game fun — but also makes it harder to stop.
- Creative Tools: Apps that generate art, stories, or music with a few clicks can feel exciting. But sometimes, they skip over the creative process — the experimenting, trying, failing, and learning that really helps imagination grow.

What's great about AI in play

AI can actually bring some really helpful things into your child's playtime — especially when used in thoughtful ways. Here's how:

- It makes things more accessible: For kids who find traditional play tricky — maybe because of speech delays, learning differences, or social challenges — smart toys and games can feel easier to connect with. They respond quickly, don't judge, and often "get" what your child is trying to do.
- It sparks ideas: Sometimes kids just need a little nudge to get their imagination going. AI tools can help with that — offering a prompt, suggesting a twist in a story, or creating a cool image they hadn't thought of.
- It keeps things engaging: Many AI-powered games and apps adjust on the fly. If something's too hard or too easy, the game changes pace. That can help kids stay motivated without getting bored or frustrated.

What to watch out for

While AI can add fun and flexibility to play, there are a few things to keep in mind so it doesn't take over the creative process:

- Too much passive play: If a toy always leads the story or solves the puzzle, kids don't get to flex their own imagination. That means fewer chances to come up with ideas, make mistakes, and build confidence in their own thinking.
- It's designed to keep them hooked: Let's be honest — not every app or game is designed with your child's best interests at heart. Some use AI to keep kids engaged for longer, not to help them grow. The flashy features can be fun, but they're often meant to keep kids clicking rather than creating.
- Privacy matters: Some smart toys and apps collect info — like what your child says or how they play. That's why it's

important to check the settings, read the fine print (as best you can), and choose tech that respects your family's privacy.

What this means for parents

You don't have to avoid AI toys or apps altogether — just stay curious about how they're being used.

Try asking yourself:

"Is this helping my child bring their own ideas to life — or is it doing all the work?"

"Does it leave space for creativity, or always steer them in one direction?"

"Is this part of a healthy mix — or replacing the joy of offline, messy, imaginative play?"

A good rule of thumb?

If the toy or app suddenly disappeared, could your child still keep playing — making up stories, drawing pictures, or inventing games on their own?

If the answer's yes, then you're probably in a good spot.

When AI helps — and when it hurts creativity

At first, AI can feel like a creative superpower. It can draw pictures, make up stories, suggest music — even build entire fantasy worlds in a few clicks. For kids, it's like having a magic helper in their pocket.

And used well, it can be just that — a fun, empowering way to bring their ideas to life.

How AI can spark creativity

AI can be especially helpful for kids who are just finding their creative voice or who sometimes doubt their abilities. It can take the pressure off and open the door to new ideas. For example:

- A shy child might use a story app to build a fantasy world they're too nervous to describe aloud.

- A budding artist could try out different colours and styles with an AI art tool, building confidence as they go.
- A young musician might use an AI music app to remix beats or experiment with lyrics, just to see what clicks.

These tools can help kids get started, beat creative blocks, and feel more excited about expressing themselves.

The catch: skipping the struggle

AI can make things easier — and that's part of the appeal. But sometimes, "easy" isn't what helps kids grow.

Think about the joy of figuring something out after a few tries — building a tricky LEGO model, mixing the perfect paint colour, or finally getting a story just the way they imagined. That struggle is where creativity really develops. It's how kids learn to problem-solve, stay curious, and feel proud of what they've made.

When AI jumps in too soon — finishing the drawing, fixing the story, or offering instant answers — it can take away those valuable moments. Over time, some kids might start doubting their own ideas or feel like the tool is the creative one, not them.

That's why it's so important to balance AI with good old-fashioned making, imagining, and trying again. The magic isn't just in the end result — it's in how they got there.

Finding the right balance

You don't have to avoid AI altogether — it's really about how it's used. The goal is to keep your child's imagination at the centre of the experience, not sidelined by the tool.

You might ask yourself:

"Is this helping them explore new ideas — or just speeding things up?"

"Are they excited about what they created, or mostly wowed by what the tool did?"

"Would they still enjoy doing this if the AI wasn't part of it?"

Creative confidence doesn't come from polished results — it grows from trying, experimenting, and putting their own spin on things. Even when it's messy or slow, that's where the real magic happens.

Imagination vs imitation — why original thought still matters

Today's kids have amazing tools at their fingertips—AI that can whip up a story, a song, or an image in seconds. But there's a big distinction between creating something and generating something. In a world full of AI outputs, it's more important than ever to help children treasure their own unique ideas.

AI can imitate—but it can't imagine

AI is great at imitation. It can remix styles and patterns into something impressive-looking. But it doesn't ask "what if," make play out of boredom, or paint simply because it feels fun. That spark—that kind of imagination—is something only humans, especially kids, bring.

Why That Matters

When AI does too much of the thinking, kids might:

Creativity is like a muscle. If it doesn't get used, it can shrink. Those quirky, messy, gloriously original ideas aren't just for show—they're how children learn, grow confidence, and make sense of their world.

How to keep their imagination active

Here are some simple ways to nurture originality:

- Celebrate the wacky ideas.

  "Wow—that's a wild twist. I love how your mind works!"

- Ask open-ended questions.

  Instead of "Want the AI to finish this?" say, "What happens next in your story?"

- Let boredom be the launchpad.

  Some of the best creative ideas start when kids have a minute of nothing—and let their minds wander.

- Let AI be a helper, not the creator.

  Use it for color palettes or word ideas—not to do everything.

Remember: creativity isn't about getting it "right." It's about making something that only you could think of. That's something no AI can ever imitate.

Screen-free creativity still matters

These days, creativity often comes with a screen — whether it's drawing on a tablet, making videos, or using AI to write stories. And yes, that kind of play can be fun and inspiring. But there's something special — and essential — about the kind of imagination that comes from building a fort out of couch cushions or pretending a stick is a magic wand.

Even in an AI-powered world, old-school, unstructured play still plays a big role in how kids grow, learn, and express themselves.

Why screen-free play is so powerful

When kids play without screens, they're not just having fun — they're building life skills in the most natural way possible. Here's what happens:

- They make the rules — and change them as the story evolves.
- They process emotions — acting out big feelings through characters or games.
- They problem-solve on the spot — using whatever's around (a blanket becomes a cape, a cardboard box becomes a spaceship).
- They practice patience and teamwork — learning how to share, take turns, and handle frustration.

This kind of play strengthens creativity, emotional resilience, and flexible thinking — qualities that matter just as much as any digital skill.

But aren't screens just the norm now?

Absolutely — screens are everywhere, and many kids love them. Some screen-based play is great. It can connect kids with new ideas, cultures, and creative tools.

But if screen time becomes the only kind of play, children may miss out on:

- Physical movement
- Daydreaming and quiet thinking
- The experience of figuring things out for themselves
- Moments of boredom that often spark the best ideas

Screens don't have to be off limits. They just need to be balanced with other ways to play, imagine, and create.

Simple ideas to encourage screen-free creativity

You don't need fancy toys or hours of free time. Just a little space and permission to be messy, silly, or bored. Try:

- Leaving out open-ended supplies like string, tape, blocks, or dress-ups and seeing what your child invents
- Turning the living room into a stage or a fort
- Starting a story with one line: "Once there was a giant who only ate jellybeans..." and letting your child take it from there
- Pairing AI tools with hands-on creativity: "You made that picture with the app — what would it look like if you painted it yourself?"

For teens and tweens, too

Older kids need this kind of creative space just as much — even if it looks different. Think:

- Journaling or sketching just for fun
- Cooking a meal from scratch
- Playing an instrument or writing a song

- Taking on a creative challenge, like designing a board game or solving a real-world problem

They might not call it "play," but the benefits are the same: confidence, expression, and a sense of doing something real.

Screen-free play doesn't have to be perfect or planned. It just needs to happen often enough to remind kids that they are the storytellers, builders, and creators — not just the users of someone else's app.

How AI might shape creative careers

When we picture the future of work, our minds often go to tech-focused jobs like coding or engineering. But here's the exciting part: AI is also reshaping creative fields — from music and design to writing, film, and fashion. That opens up new opportunities for our kids, and it's worth exploring together.

The creative industries are changing

AI is already helping professionals bring ideas to life faster:

- Drafting scripts or ad copy
- Composing music or remixing tracks
- Generating artwork and design mockups
- Animating short films or prototyping games

It's not just large studios using this tech—freelancers, artists, and small businesses are too. In fact, a teenager with a laptop might now create work that once required a whole production team.

What this means for the next generation

If your child dreams of drawing, writing, performing, or filmmaking, AI tools might soon be part of their creative toolkit. Here's how that could look:

- Quicker output — they'll be able to produce more, but may feel pressure to stand out.

- More solo paths — self-publishing, indie art, DIY music + marketing, but also more responsibility.
- New hybrid roles — like prompt design, ethical AI filtering, or creating interactive stories with AI guides.

But what about originality?

A big concern many parents and creatives share is this: if a machine can write a song or paint a picture, where does that leave human creativity?

Here's the truth: AI can replicate, but it can't feel. It doesn't understand joy, sadness, humour, or what makes a moment truly meaningful. It works by remixing patterns — not creating from lived experience. And that's exactly where people shine.

In a world flooded with AI-generated content, human originality becomes even more valuable — especially when it's rooted in real emotions, personal stories, and thoughtful values.

It's also worth noting: copyright is still a murky area. Some AI-generated work borrows from existing artists' styles without permission, raising big questions about ownership and ethics. Teaching kids to respect creative work — both their own and others' — is more important than ever.

What you can say to your child

If your child is into music, gaming, YouTube, fashion, or writing, you might be wondering how AI fits into that future. The goal isn't to worry them — it's to encourage them in a grounded, empowering way.

Try saying things like:

"Tech can be a really helpful tool — but your creativity is what makes something special."

"Learning how these tools work means you can use them smartly — not be replaced by them."

"People still care about stories, feelings, and ideas that come from the heart. That's what stands out."

These kinds of conversations show kids they're not just along for the ride — they have a role to play in shaping the creative future.

Raising creative thinkers in a tech-heavy world

Play, imagination, and creativity aren't just extras in childhood. They're the foundation for learning, emotional well-being, and — yes — future success in a world where AI is always getting smarter.

Whether your child is building LEGO kingdoms, writing funny songs, designing digital comics, or planning imaginary adventures for a pretend YouTube show, their creativity is the source of everything real. It's how they express themselves. It's where their ideas first take shape. And it's the fuel that helps them stay curious, resilient, and hopeful — even as the tools around them keep evolving.

Make no mistake: AI will continue to change how kids create, explore, and share. But that doesn't mean childhood is disappearing. It means parents have an important job now more than ever: to keep space open for imaginative play, to cheer originality, to guide their kids as they learn to mix timeless creativity with powerful new tools.

As parents, we bring something powerful — not answers, but awareness. We're the ones who notice what makes our child's eyes light up. Who celebrate the spark when they create something all their own. Who gently steer them toward making, not just consuming. And in doing that, we give them a moment to pause, explore, and say, "This came from me."

Whatever your child grows up to be — an artist, a coder, a writer, a teacher, or an entirely new career we've never imagined — this stays the same:

Creativity is their superpower. And your curiosity, encouragement, and presence make it shine.

Chapter Seven: How Australia Is Responding to AI

What kind of future are we building — and who gets a say?

It's one thing to understand how AI works. It's another to shape how it's used — in our schools, services, and daily lives. That's where national policy comes in.

This chapter looks at what Australia is doing to guide AI in ways that reflect our values: safety, fairness, inclusion. It's not just about high-level plans or tech jargon — it's about protecting people, especially children, as AI becomes part of everyday life.

We'll explore how government, industry, and communities are working (and sometimes struggling) to set boundaries, ask hard questions, and build trust in a fast-moving digital world. We'll also look at why even the world's top AI experts are warning us to tread carefully — and what families need to know now.

Because the future isn't something we wait for. It's something we shape — together.

Why a national strategy is important

It's easy to think of AI as something that's happening "somewhere else" — in big tech labs overseas or in boardrooms we'll never sit in. But the truth is, AI is already part of daily life here in Australia. And often, we don't even realise it.

Maybe your child uses a learning app that adapts as they go. Maybe you've had a Medicare update, or a Centrelink payment processed automatically. Maybe your teen's social feed seems to read their mind. AI is behind the scenes in all of it.

That's why it matters how we, as a country, choose to guide this technology.

And we are guiding it. Australia has a national Digital and Data Strategy (https://www.dataanddigital.gov.au/) — a kind of big-picture plan for how new tech, including AI, should work for people. The goal? Systems that are safe, fair, and built on trust. That means protecting kids' rights, making sure all families have fair access, and making space for communities to be part of the conversation.

This chapter isn't just about policies on paper. It's about the kind of digital future we're creating — and the values we want to hold onto as AI becomes part of more decisions.

Because when tech moves fast, someone needs to ask the big questions:

Is this fair?

Who's deciding how it works?

What happens when it goes wrong?

And how do we make sure no one's left out?

That's what a national strategy is really for. Not to stop progress, but to put the right boundaries in place — the kind that say: this is how we use AI responsibly, this is how we protect people, and this is what we care about as a society.

For families, that brings some comfort. It means the tools your child uses at school or at home aren't just about what's easy or efficient — they're being shaped by policies that try to put people first.

And here's something else. Some of the very people building AI — the insiders, the inventors — are starting to speak up. They're worried. Not just about what AI can do now, but where it's heading. Too powerful, too fast, and too hard to stop if we don't get the rules right early. That's why this work matters — not later, but now.

So yes, AI is already here. But so are the conversations, the questions, and the guardrails that can help us use it wisely.

Australia's key AI initiatives (explained simply)

We've talked about the importance of national guardrails – not just for keeping up with AI, but for making sure it works in ways that reflect our values.

So what do those guardrails actually look like?

Across government, there are now practical tools in place to help guide how AI gets designed and used – especially in areas that affect families. You might hear terms like ethics principles or voluntary safety standards. And while they can sound a bit technical, they're built around ideas that most of us already understand: fairness, accountability, and protecting people's rights.

These aren't rules for some distant future – they're steps Australia is taking now, in line with our national digital strategy, to make sure AI is safe, inclusive, and people-first.

Let's take a closer look – and unpack what these frameworks really mean for kids, parents, and communities.

The AI Ethics Principles

Australia's AI Ethics Principles were introduced to help guide how artificial intelligence is used across sectors – from government services to education to business.

While the language might sound formal, these principles are grounded in ideas most of us already teach our kids: fairness, honesty, responsibility, and respect.

They're also part of a bigger effort to build public trust – a key goal in Australia's national approach to digital systems. The idea is simple: if AI is going to be part of daily life, it needs to work in ways that genuinely support people.

Here's how all eight principles translate into everyday terms — especially for families:

1. **Human, social and environmental wellbeing**

   AI should help people, not harm them — and it should support communities, society, and the environment.

   Example: A local council uses AI to manage traffic lights to reduce emissions and improve safety near schools. That's AI supporting wellbeing.

2. **Human-centred values**

   Technology should respect human rights, dignity, and autonomy. In short: it should work for people.

   Example: A reading app for children lets them choose their pace and provides support without judgment, respecting their individual needs.

3. **Fairness**

   No one should be treated unfairly by an algorithm — especially not because of things like race, gender, disability, or postcode.

   Example: An AI tool used to screen scholarship applicants should be checked to ensure it doesn't unintentionally favour students from certain suburbs or schools.

4. **Privacy protection and security**

   Your personal data — or your child's — should be handled safely, securely, and respectfully.

   Example: If a school uses AI to track attendance, it should clearly explain how data is stored and who has access.

5. **Reliability and safety**

   AI systems should work as intended. If they're used in schools, hospitals, or support services, they need to be trustworthy and well-tested.

   Example: An AI maths tutor should consistently provide accurate feedback — and flag when it's unsure instead of guessing.

6. **Transparency and explainability**

   If a decision is made by AI, people should know about it — and be able to understand how and why it happened.

   Example: If your child is recommended for a learning extension program by an AI, the school should explain how that decision was made.

7. **Contestability**

   If you disagree with a decision made by AI (like being denied support or being graded unfairly), there should be a clear way to question it.

   Example: If a family is affected by an automated decision about government benefits, there should be a simple way to challenge or appeal it.

8. **Accountability**

   There must be a person — not just a system — responsible for AI decisions and outcomes.

   Example: If an AI-powered chatbot gives misleading medical advice, someone should be responsible for reviewing and correcting it.

These principles aren't about slowing down technology. They're about making sure it works for everyone — especially kids and families who may not see how the decisions are made behind the scenes.

Voluntary AI Safety Standards: what good AIlooks like

In 2024, Australia released a set of 10 voluntary safety standards to help guide how AI is used in real-life settings — from schools and government services to the apps our kids use at home.

Think of these like a checklist for responsible AI — not legal rules (yet), but clear expectations. They're part of a wider push to make sure digital systems are designed with care, not just convenience.

These standards build on the ethics principles we just looked at. They zoom in on what "good AI" looks like in practice — especially when real people, and especially children, are affected by the outcomes.

The idea is simple: if an AI tool is being used to teach a child, assess a welfare claim, or personalise support, it should be safe, explainable, and easy to challenge if something goes wrong.

Here's what those 10 standards mean in everyday terms — and how they might look in the family context

1. **Accountability**

   Someone must be responsible for what AI does — it can't be left on autopilot.

   Example: If your child's learning platform gets something wrong, a teacher or staff member should be able to explain and fix it.

2. **Risk management**

   Organisations should look for risks early — and have a plan to handle them.

   Example: A school trialling an AI chatbot should think through how it might be misused and build in safety checks.

3. **Data governance**

   Personal information — like your child's name, age, or learning habits — must be handled with care.

Example: A reading app should clearly explain what data it collects, how it's stored, and who sees it.

4. **Testing and monitoring**

AI tools should be tested before launch — and checked regularly to make sure they're still working properly.

Example: A maths app should be reviewed often to make sure it's giving accurate, helpful feedback.

5. **Human oversight**

People need to be able to step in. AI should never have the final say on something important.

Example: If an automated system flags a student for truancy, a teacher should review it before any action is taken.

6. **Transparency**

Families should know when AI is being used — and what it's doing.

Example: If a school uses AI to personalise newsletters or reports, it should be open about how those choices are made.

7. **Contestability**

There should always be a way to question or appeal AI-driven decisions.

Example: If a child is denied access to a support program based on an AI assessment, parents should have a way to challenge it.

8. **Supply chain transparency**

Schools and services should know where their AI tools come from — and what they do.

Example: If a school uses a third-party AI tool from overseas, it should be able to explain what it collects and how it works.

9. **Record keeping**

Organisations should keep clear records of how AI decisions are made.

Example: If an AI grading tool makes a recommendation, there should be a log showing how it reached that result.

10. **Stakeholder engagement**

People affected by AI should be part of the conversation — before it's rolled out.

Example: If a school plans to use AI for behaviour monitoring, families should be consulted before anything starts.

These standards don't cover every possible risk — but they give schools, businesses, and government agencies a clear place to start. And they help parents know what kinds of questions to ask when AI shows up in your child's life.

They also reflect what Australians have said loud and clear in recent consultations: trust, inclusion, and transparency matter. And families want to know that AI is being used with care, not just speed.

Mandatory guardrails for high-risk AI: where the stakes are highest

While many AI tools are helpful and low-risk — like a spelling app or a chatbot that answers simple questions — others can have a big impact on people's lives. Think school assessments, job interviews, health advice, or welfare applications.

That's why Australia is now considering a more serious step: making the guardrails mandatory for high-risk AI.

These rules wouldn't apply to every app or gadget — just to AI systems used in settings where a mistake could have real consequences. The kind of situations where fairness, accuracy, and accountability aren't just "nice to have" — they're essential.

Under the current proposals, any organisation using high-risk AI would need to:

- Be open about it

  People should know when AI is involved in decisions — no guessing, no fine print.

  Example: If your teen's job application is screened by AI, they should be told upfront and given a clear explanation of how it works.

- Let people challenge it

  If someone thinks a decision is unfair or wrong, there should be a clear and easy way to ask questions or appeal.

  Example: If a student is denied extra support because of an AI prediction, families should be able to request a human review.

- Take responsibility for outcomes

  If something goes wrong — if an AI tool causes harm or makes a biased decision — someone must be accountable.

  Example: If an AI-powered system misclassifies a learning disability, the school or service provider must step in and make it right.

These proposals reflect a bigger shift in thinking: recognising that some decisions are just too important to leave entirely to machines.

The government is currently reviewing where existing laws fall short — and where new rules might be needed to fill the gaps. It's not just about future-proofing policy — it's about building trust, one safeguard at a time.

For families, this means more peace of mind. Knowing that if AI is involved in something that really matters, there are safety nets in place — not just in theory, but in practice.

The AI Impact Navigator: measuring what really matters

Not all AI impacts show up in performance metrics or technical benchmarks. Sometimes the real effects — especially on people — are harder to measure.

That's why the National AI Centre created the AI Impact Navigator. It's a self-assessment tool for schools, government agencies, and organisations to step back and ask: How is this technology actually affecting people's lives?

It's designed to shift the conversation from "Does it work?" to "Is it working well for the community it serves?"

The Navigator prompts decision-makers to reflect on questions that matter — not just to developers, but to families, educators, and everyday users. Questions like:

- Is this AI helping the community?

  Does it support equity, wellbeing, and social good — or leave some groups behind?

- Are people being treated fairly?

  Is the technology making things more accessible and inclusive, or unintentionally reinforcing old biases?

- Are staff supported as tech changes?

  Are educators, admin teams, or health workers being trained and involved — or just expected to adapt?

- Is the system transparent?

  Can families understand what the AI is doing? Are they informed, or left in the dark?

Why it matters for families

For parents, tools like the AI Impact Navigator are a welcome signal: that AI isn't just being judged on speed or efficiency, but on how it affects real people in everyday settings.

It helps organisations move beyond convenience — and focus on fairness, clarity, and care.

And while it doesn't replace regulation, it fills an important gap. It encourages thoughtful design from the outset, rather than waiting to measure impact after the fact.

Because when technology is designed with people in mind from the start, families can feel more confident that it's working with them — not just around them.

Where Australia stands

Australia has strong foundations when it comes to AI — and importantly, they're built around people, not just tech.

We've got nationally agreed AI Ethics Principles.

There's growing focus on responsible use in education and government.

Tools like the AI Impact Navigator help assess real-world effects, not just performance stats.

And we're now in the middle of a public consultation process to turn some voluntary standards into mandatory rules — especially where the stakes are highest.

What's still evolving is how we bring all of this into everyday life — into classrooms, public services, and the kinds of tools families use at home. Other countries show that things like public awareness, inclusion, and plain-language tools can make a big difference.

And in 2025, Australia's conversation about AI is definitely picking up pace.

Here's what's been happening lately — and why it matters for families:

- Teen Safety Online

New legislation in late 2025 will require social media platforms to block under-16s — using AI-based age checks that are minimally invasive. Instead of scanning IDs, the focus is on behavioural signals. It's a sign that Australia is trying to balance protection with privacy — especially for young people online. *Source: Reuters*

- Public Support for Stronger Rules

A recent poll shows that most Australians — including parents — want clear, enforceable guardrails. 61% favour a balanced approach to AI. 64% say regulation should take priority, even if it means slower innovation. That sends a strong message: safety and fairness come first. *Source: The Australian*

- Fairness and Inclusion Still a Concern

Australia's Human Rights Commissioner recently warned that without the right checks in place, AI could unintentionally reinforce bias — including racism and sexism. The push is for more transparency, local data, and human oversight so families don't get left out or left behind. *Source: The Guardian*

- Pressure to Catch Up on Infrastructure

Telstra's CEO has highlighted the need for stronger digital infrastructure and clearer, faster regulation to help Australia keep pace with global tech shifts. It's not just about big business — it's about making sure our systems work well and fairly for everyone. *Source: National Press Club presentation 2025*

- Copyright and Creators

There's also growing debate about how AI tools use creative content. The ACTU and Tech Council (with members like Google and Microsoft) are exploring a model to compensate creators whose work trains AI. But authors and musicians are pushing back — arguing that any deal must respect Australian copyright laws, not just big tech interests. *Source: ABC News*

- Regulation Under Review

The government is weighing up whether to introduce a new "AI Act" — or update existing laws to better cover emerging risks. For now, they're starting with a regulation gap analysis. In plain terms: they're mapping what we already have, what's missing, and what needs fixing. *Source: Digital.gov.au*

How Australia compares globally

Australia isn't the only country grappling with how to guide AI responsibly. Around the world, governments are experimenting with different approaches — especially in schools, healthcare, and public services that families rely on.

Here are some highlights — and what we can take away from them:

Finland: teaching everyone about AI

Finland became a global leader in digital education by doing something simple but powerful: they made AI education available to everyone. Their free "Elements of AI" course has been taken by more than a million people — students, parents, grandparents, and professionals. *Source: University of Helsinki*

Why it matters: Finland shows that AI literacy isn't just for tech experts. It can be part of everyday learning, helping communities feel informed instead of left behind.

Singapore: safe, inclusive AI design

Singapore's national AI strategy puts safety and inclusion front and centre. They work closely with schools to trial AI tools responsibly and have special protections in place for children and the elderly. *Source: Smart Nation Singapore*

Why it matters: Singapore proves that AI policy isn't just about the economy — it can be designed to protect the most vulnerable, and tested in real-life settings before being scaled up.

European Union: The AI Act

The EU's Artificial Intelligence Act officially became law in 2024. It uses a risk-based approach: low-risk AI is lightly regulated, while high-risk applications — like healthcare, education, and biometric identification — face strict transparency and safety requirements. *Source: European Parliament*

Why it matters: The EU is showing what it looks like to put clear, enforceable guardrails in place. For Australian families, it's a reminder that we could see similar risk-based rules here in future.

United Kingdom: building trust through regulation

The UK is preparing a more comprehensive AI bill after delaying earlier plans. The new bill is expected to combine innovation with stronger safeguards — especially around testing large AI models and protecting people in employment and healthcare. *Source: The Guardian*

Why it matters: The UK is working to balance growth and guardrails — and putting trust at the centre. That's the same tension Australia is navigating now.

United States: big tech, patchy rules

The US is home to many of the world's biggest AI companies — from educational tools to social media platforms. Much of the AI your child interacts with (YouTube's recommendations, homework helpers, voice assistants) likely comes from the US.

Earlier attempts at a national rulebook stalled. In 2025, Congress couldn't agree on strong federal legislation to set clear standards for AI. That left a big gap.

Into that gap stepped the states. California, New York, Texas and many others are now writing their own laws. These cover things like children's online safety, how employers use AI to screen job applications, and when companies must explain automated decisions. Some states are even banning harmful uses of AI, like non-consensual deepfake images. *Source: The Conversation*

Why it matters: This patchwork means there's no single standard shaping the apps our kids use — protections vary depending on where the company is based. That's why Australia needs its own clear guardrails. We can't assume US rules will automatically protect Australian families, because in many cases, they don't exist yet.

Canada: building ethics into everyday systems

Canada has taken a proactive approach to AI, especially in government services that families rely on. Before an agency can roll out a new AI system, it has to complete an Algorithmic Impact Assessment. That's just a checklist of questions about risk, fairness, and transparency — asked before the technology goes live.

Canada is also putting energy into digital inclusion, making sure people in remote areas and underrepresented communities have a voice in how AI is designed and used. *Source: Government of Canada*

Why it matters: Canada is showing that ethics don't have to be an afterthought — they can be built in from the start. For families, that means more confidence that the systems supporting kids, schools, or healthcare are being tested for fairness and explained in plain language.

New Zealand: wellbeing first, with a cultural lens

New Zealand takes a unique approach to AI. Their government guidance is clear: technology should respect community values, cultural safety, and Māori data sovereignty — meaning Indigenous communities have a say in how their data is used.

The focus is on responsible, transparent AI that serves people first. Agencies are expected to check for fairness, explain decisions clearly, and keep human oversight in place. *Source: NZ Government – AI Guidance*

Why it matters: New Zealand reminds us that technology isn't only about speed or efficiency. It should reflect the values of the community it serves — and protect people's rights and wellbeing along the way.

If the creators are worried, shouldn't we be too?

It feels like everywhere you look in the news, AI experts are sounding the alarm. And these aren't just critics from the sidelines — they include some of the very people who built today's most powerful AI systems. Instead of only celebrating progress, many are warning about what could go wrong if we're not careful. At times, the headlines even carry a sense of doom and gloom — predictions about machines taking over or outsmarting humans.

But underneath the dramatic language is a simple message: AI is moving fast, and we need to make sure our guardrails keep up.

Some of the most respected voices in AI have been raising these concerns:

- **Geoffrey Hinton** — often called the "godfather of AI" — left Google in 2023 so he could speak freely about his worries. He has warned that AI could become smarter than us sooner than expected, and that we don't yet know how to keep it under control.
- **Mustafa Suleyman**, co-founder of DeepMind and author of The Coming Wave, argues that AI could be as transformative as electricity — but also as dangerous as nuclear weapons if left unchecked. His point is that the stakes are global: just as the world had to come together to manage nuclear risks, we'll need similar cooperation and safeguards for AI.
- **Roman Yampolskiy**, an AI safety researcher, writes about "superintelligence" — the possibility of AI eventually

surpassing human intelligence in ways that are hard to predict or manage. His message is simple: better to plan for extreme risks now than regret it later.

- **Stuart Russell,** author of Human Compatible, has become one of the clearest voices on AI caution. Back in 2023, he signed the now-famous "pause letter," which called for a temporary halt to AI development because things seemed to be moving too fast. His concern then was runaway progress without enough safeguards. Now, only a couple of years later, his worry has shifted in the opposite direction: that the industry is overheating on sky-high expectations that could collapse suddenly, just like the "AI winter" of the 1980s. Russell's perspective carries extra weight because he has seen both sides — the risks of unchecked speed, and the risks of inflated hype.
- **Yoshua Bengio**, one of the pioneers of deep learning, has shifted much of his work toward ethics and safety. He emphasises fairness and accountability, especially when AI is used in education, health, or government.
- **Yann LeCun**, Meta's chief AI scientist, has long said today's AI tools have limits. He reminds people that large language models aren't magic, and that progress may slow unless new breakthroughs emerge.

It's not just the experts speaking up — the news is full of stories that make parents stop and think.

In the US, some parents have stood before Congress to share heartbreaking accounts of AI chatbots encouraging their teenagers to self-harm. That's now led to regulators asking tough questions of big tech companies: are they really doing enough to keep kids safe? Closer to home, Australia's eSafety Commissioner has raised the alarm about deepfake abuse in schools — with students using AI to create fake, sometimes explicit, images of their classmates. For the families caught up in it, the harm is immediate and very real.

Then there are the new AI "companions" and toys hitting the shelves. They talk back, respond with empathy, even sound like they care. Some kids love them — but some parents describe the experience as unsettling, even "creepy," worrying that children might mistake programmed affection for the real thing. And layered on top of all that is a new generation of scams, powered by AI voices and messages so convincing that even adults are getting fooled.

Put together, it's a lot. And it's a reminder that AI isn't just a "future" issue or some abstract risk. It's already in our homes, schools, and communities. Families can do their part — teaching kids to ask questions, think critically, and be cautious — but there's only so much parents can do alone.

That's why we need stronger guardrails. It shouldn't be up to mums and dads to wonder whether a school app, a toy, or a social platform has been tested for safety. That responsibility should sit with the companies building these tools — and with the policymakers setting the rules. Because AI is already part of our children's lives.

The real question is: are we shaping it on our terms, or theirs?

Where this leaves us

Looking across these countries, one thing is clear: there's no single "right way" to manage AI — but there is a shared goal. Whether it's Finland teaching digital confidence, Singapore trialling safe tech in schools, or New Zealand protecting cultural values, each country is finding its own path to responsible AI.

Australia is part of that global effort. And while we may be quieter than some tech giants, our focus on ethics, inclusion, and real-world impact gives us a strong foundation.

For families, it's a reminder that AI isn't just a global story — it's a local one. The choices we make here, in our schools, homes, and communities, shape how this technology fits into everyday Australian life.

At first glance, all this talk of frameworks, principles, and policy might feel a little abstract — like something best left to government departments or tech companies. But when you dig deeper, these efforts are really about something we all understand as parents: creating safe, fair, and supportive environments for our kids.

Whether it's the classroom, the sports field, or the dinner table, we spend a lot of time helping our children navigate decisions. We teach them to ask questions, to notice when something feels off, and to look out for others — not just themselves.

Now, those same values are being applied to the technology that increasingly shapes their world.

Because the truth is, AI is already making decisions that affect families. It might be helping a school identify which students need extra support. It could be ranking a job application your teen submits online. It might even be flagging behavioural trends or shaping the ads your child sees on their screen.

And while some of that can be helpful — saving time, offering personalisation, spotting patterns — it can also feel invisible, hard to challenge, or unfair. That's where policy steps in. The ethics principles, voluntary standards, and proposed guardrails are about making sure humans stay in the loop — and that we have the right to understand, question, and influence how these tools are used.

They also help shift the conversation from "What is this AI doing?" to something more practical for families: "Who is it serving?" That shift matters, because it prompts us to ask the kinds of questions that really cut through:

- Who built this tool, and what assumptions shaped it?

- Is it helping all students equally, or leaving some behind?
- What happens if it makes a mistake?

And maybe the biggest question of all: What kind of future are we shaping — not just with AI, but alongside it?

Australia may not have the flashiest AI systems, but we have something just as powerful: a national conversation that's grounded in values. It's early days, and there's still work to do — especially in making these standards visible and accessible to everyday families. But the foundations are there.

By staying informed, asking questions, and bringing these conversations into schools, workplaces, and communities, parents can play a quiet but crucial role. Not just in protecting children, but in helping technology reflect the best of who we are — thoughtful, fair, and human.

Chapter Eight: Equity, Access & the Digital Divide

We all want our kids to have the best possible start — to learn, connect, and grow in a world full of opportunity. But what happens when some families have better access to that world than others? When the tools meant to help only work well for certain people — or miss others entirely?

In a world shaped by AI, the question of who's included — and who's left behind — matters more than ever. This chapter explores what that looks like in real life, why it's not just about Wi-Fi or devices, and how we can work toward a future where all families feel seen, supported, and empowered.

What we mean by the digital divide

When people talk about the "digital divide," it often starts with a basic question: Who has internet access — and who doesn't? That's still a big issue in Australia. But in the age of AI, the divide runs deeper. It's not just about getting online — it's about who is included in the design, development, and benefits of technology, and who is being left behind.

For families, this divide can show up in everyday ways:

- A student in a remote town trying to complete homework with patchy internet and limited data.
- A parent struggling to use a school app that assumes everyone speaks English fluently or has the latest phone.
- A child whose learning style doesn't "fit" the assumptions of an AI tool, and gets flagged unfairly.

These aren't isolated problems — they reflect broader patterns.

In Australia, one of the tools we use to track this is the Australian Digital Inclusion Index.

But it doesn't just ask, "Do you have the internet?" It goes deeper:

- Can you access it reliably?
- Can you afford the data, devices, and upgrades?
- Do you feel confident using it — or are you just muddling through?

What it shows, year after year, is this: digital access and confidence are not evenly spread.

Rural communities, First Nations families, older Australians, and low-income households face more barriers — not just in getting online, but in knowing what to do once they're there.

That gap becomes even more important when AI gets involved. Because unlike other tools, AI can quietly make decisions about your child's education, health support, or service access — sometimes without you even knowing it. And if the people building those tools haven't thought about your child's context, culture, language, or learning style... the system might not work fairly for them.

In Australia, one of the tools we use to track this is the Australian Digital Inclusion Index.

But it doesn't just ask, "Do you have the internet?" It goes deeper:

- Can you access it reliably?
- Can you afford the data, devices, and upgrades?
- Do you feel confident using it — or are you just muddling through?

What it shows, year after year, is this: digital access and confidence are not evenly spread.

Rural communities, First Nations families, older Australians, and low-income households face more barriers — not just in getting online, but in knowing what to do once they're there.

That gap becomes even more important when AI gets involved. Because unlike other tools, AI can quietly make decisions about your child's education, health support, or service access — sometimes without you even knowing it. And if the people building those tools haven't thought about your child's context, culture, language, or learning style... the system might not work fairly for them.

In this book, when we talk about the digital divide, we're not just talking about slow Wi-Fi or who owns a laptop.

We're talking about something bigger:

Fairness. Representation. Opportunity.

It's about who gets to participate fully in the digital world — and who gets left behind because no one asked the right questions when the tools were built.

As parents, we teach our kids to notice when something feels unfair.

We tell them, "Speak up if something doesn't make sense," and "Look out for others, not just yourself."

Now, we need to apply that same thinking to the technology around them.

Because AI is already shaping what our kids see online, how they learn, and even how support services are delivered.

So we have to ask:

Is this working for all families — or just the ones who already have access, language, and confidence?

And more importantly:

How can we make sure everyone's included in what comes next?

Who's most affected and why it matters

The digital divide doesn't impact everyone in the same way. Some families can quickly pick up new tech tools or jump on the latest app. But for others, it's a different story — and not because they're lazy or uninterested.

It's often because the system wasn't built with their lives in mind.

Here are some of the groups most affected — and why it's something we all need to care about:

Families in rural and remote areas

Connectivity remains a major issue outside the major cities. Even where internet is available, speeds can be unreliable or expensive. For families in regional and remote areas, this limits access to digital learning, telehealth, and other AI-enhanced services.

Why it matters:

Think about what happens when:

- A school uses an AI-powered learning app that needs fast internet — and your child keeps getting dropped out of the lesson.
- A tutoring platform relies on video, but the connection keeps freezing.
- A healthcare service offers online appointments, but your family can't log in without using up all the month's data.

None of these issues are your child's fault — but they can still feel like they're falling behind. That's frustrating for them, and stressful for parents trying to support them.

During COVID-19 lockdowns, many students in remote Northern Territory communities couldn't fully take part in online learning. Some didn't have enough devices to go around at home, and others were stuck with slow or unreliable internet.

Teachers tried their best to work around it — but the reality was clear: our digital infrastructure doesn't reach everyone equally.

Low-income families

When money is tight, tech access often becomes a juggling act. Families might share one device across multiple kids, rely on public Wi-Fi, or skip online programs that cost extra. It's not about lack of interest — it's about making ends meet.

Why it matters:

AI-powered learning tools often assume that every child has a device, stable internet, and a quiet place to learn. But in many homes, that's far from reality. If an AI system uses login times or quiz completion to assess progress, it might misread the situation — flagging a student as struggling when really, they just can't get online consistently. In those moments, AI can end up measuring access — not ability.

"Digital inclusion for all public school students" — AEU Report

Commissioned by the Australian Education Union, this study surfaced real stories from disadvantaged students — including those who relied on public Wi-Fi or shared devices during COVID-19 online learning. The findings underscore how lack of access affects learning outcomes, and how tools (including AI-supported platforms) can misinterpret these gaps.

First nations communities

For many First Nations families, the digital divide isn't just about internet access.

It runs deeper — into questions of language, trust, and whether technology understands the communities it's meant to serve.

Even when digital tools are available, they're not always built with First Nations voices in mind. Some platforms don't support Aboriginal and Torres Strait Islander languages. Others don't reflect local knowledge systems, cultural practices, or ways of learning that are central to community life.

For many parents and carers, this can make technology feel more like a barrier than a bridge — especially when it doesn't feel respectful, safe, or responsive to their children's realities.

Why it matters:

If AI is used in schools, health services, or Centrelink-style programs, it needs to be built with communities — not just for them. That means listening to First Nations voices, involving Elders and community leaders, and making sure the systems reflect local knowledge, values, and ways of living.

When that doesn't happen, the risks are real. AI can misunderstand how people speak, make unfair assumptions based on incomplete data, or flag families for the "wrong" reasons — simply because it wasn't designed with their context in mind.

This isn't just about bad tech. It's about the trust families place in systems — and what happens when that trust is broken.

In some parts of Australia, technology struggles to understand Aboriginal English. A recent project by Google and the University of Western Australia is working to fix this by teaching AI how to recognise Aboriginal English more accurately. *Source: sbs.com.au* Research shows that when systems don't understand the way First Nations people speak at home, it can lead to confusion, frustration, and mistrust. Imagine trying to use a chatbot for help and it doesn't even understand your words—let alone your needs.

This isn't just a tech flaw—it's a sign that not every voice is being heard as generative tools become more common.

People with disabilities or neurodivergent learners

Technology — especially AI — can be a game changer for people with disability. Tools like voice assistants, captions, and personalised learning apps can make things more accessible and flexible. But only if they're built to include everyone.

Why it matters:

Some AI systems aren't designed to recognise different learning styles or ways of interacting. That means a child who thinks or learns differently might be flagged as "off track" — not because they're struggling, but because the system doesn't understand them. And for adults, applying for support or using online services can be stressful if the tech is confusing, fast-moving, or hard to navigate.

Research from organisations like CYDA (Children and Young People with Disability Australia) and evaluations of the NDIS show how digital barriers — like hard-to-use forms or rigid systems — can limit access to support. These aren't personal failings. They're design problems.

Some families shared that applying for NDIS support online felt like running an obstacle course — too many steps, unclear language, and no room to explain things in a different way. When tech doesn't work for people with different needs, it's not inclusive — and that has real-life consequences.

Linguistically diverse families

Families who speak languages other than English often face extra challenges when using digital tools. That's because many AI systems are trained mainly on English — especially standard grammar and vocabulary. So they may struggle with different dialects, cultural phrasing, or even just everyday ways people speak at home.

Why it matters:

When an AI system — like a chatbot or translation tool — can't understand your words properly, it may also get your needs wrong. That could mean being sent to the wrong support, being misunderstood in a school setting, or even getting incorrect advice.

This isn't just frustrating – it can affect access to important services. Research from organisations like the Scanlon Foundation and the eSafety Commissioner highlights how language barriers still impact digital confidence and outcomes across Australia.

Data from the Australian Digital Inclusion Index shows many families speaking languages other than English face lower confidence and ability in using digital tools – not because they're less capable, but because the tools and support don't match their experience.

Bias in automated systems

Even when AI tools seem "neutral," they can still carry hidden biases – especially if the data they're trained on reflects real-world inequalities. This can quietly shape everything from who gets shortlisted for a job to which families get extra support at school.

Why it Matters:

If an AI system makes decisions based on patterns in past data – like school performance, income, or postcode – it might unintentionally favour certain groups over others. That means families from less-resourced areas, or kids who learn differently, could be unfairly overlooked or flagged.

Research has raised concerns that some automated job screening tools in Australia use location data (like your postcode) as part of the selection process. That might sound harmless – but in practice, it can favour candidates from wealthier suburbs and disadvantage equally qualified applicants from low-income or diverse areas. *Source: Human Technology Institute, UTS – "Responsible AI Index" and media coverage on bias in recruitment platforms*

Across all these different groups – rural families, low-income households, First Nations communities, people with disabilities, those from diverse language backgrounds — there's a common thread: when technology isn't designed with real people in mind, it can quietly make life harder.

Often, no one sets out to exclude or disadvantage anyone. But when the people building AI systems don't reflect the full mix of communities using them, important differences get missed. And those blind spots can lead to unfair outcomes — like a student being marked as "behind" because their internet dropped out, or a parent struggling to access support because a chatbot can't understand their accent or phrasing.

These aren't just tech hiccups — they affect real lives.

That's why this conversation matters so much for families. Because it's not just about how AI works. It's about what AI affects.

It touches:

- Whether your child gets extra help or is unfairly flagged in school.
- Whether you can access services like Centrelink or NDIS smoothly — or hit roadblocks.
- Whether your teen sees job opportunities that fit them — or gets filtered out before even getting a chance.

Technology shapes opportunity. And when AI tools are built without considering difference — in culture, ability, income, language, or location — they risk reinforcing the very gaps we're trying to close.

That's why inclusion from the start isn't a "nice to have." It's essential. Because every family deserves to feel seen, supported, and respected — not left out because the system didn't know how to understand them.

What inclusive AI looks like

If all the talk about bias and exclusion feels a bit overwhelming — take heart. The good news is this: inclusive AI isn't just a dream. It's already being built, tested, and used in ways that make life easier for families.

All over the world, smart and thoughtful people — including designers, teachers, community leaders, and policy-makers — are finding better ways to make sure technology works for everyone, not just a select few.

So, what does inclusive AI actually look like in everyday life?

- It looks like a school app that translates messages into your home language — not just English.
- It looks like an AI reading tool that adapts to your child's pace and learning style, instead of labelling them as "behind."
- It looks like a government website that works just as well on a budget phone with a slow connection as it does on a brand-new laptop.
- It looks like an online form with voice support for people who have trouble typing — or simplified steps for those new to using digital services.

When AI is designed with real families in mind — with all their different needs, strengths, and challenges — the result is tech that feels like an ally, not a barrier.

This kind of AI doesn't just work better. It is better. Because it's rooted in something all parents care about: fairness, empathy, and helping people thrive.

And as we think about the role of AI in our kids' lives — at school, at home, in the services we rely on — this is the kind of future worth aiming for.

It's easy to say "make tech more inclusive," but what does that mean in real life — in our kids' classrooms, homes, and communities?

Let's break it down.

It's designed with diversity in mind — from the start

Inclusive AI doesn't wait until the end to ask, "Is this fair?" It starts by asking, "Who might this leave out?" — and then works to include them from day one.

That means making sure a wide range of people — including those who are often left out — help shape how the technology is built and tested.

What it looks like:

Imagine a reading app that isn't just trialled in top schools in big cities, but also in small rural classrooms, with families who speak different languages at home, and with kids who learn in unique ways.

The developers don't just get feedback from tech-savvy parents and confident students. They also speak with kids who use assistive technology, parents who are still learning to navigate digital tools, and educators working in low-resource areas.

Why it matters:

When AI is only tested with one kind of user in mind, it can miss important things — like how it responds to different accents, learning speeds, or tech setups. But when you include more voices from the beginning, you build a tool that's stronger, smarter, and more helpful for everyone.

It uses data responsibly — and questions its own assumptions

AI systems learn from data — and that data often reflects the real world, with all its inequalities and blind spots. Inclusive AI doesn't just take those patterns at face value. It asks hard questions: "Is this fair?" and "Who might this hurt?"

What it looks like:

Imagine a job application system that notices it's recommending more students from certain suburbs or private schools. Instead of brushing that off, the team behind it pauses, checks for bias, and updates how the system works.

They don't assume the tech is always right — they treat it like a work-in-progress that needs watching and fine-tuning.

Why it matters:

AI can unintentionally learn to copy old inequalities — like favouring wealthier areas or certain backgrounds — even if no one meant to build it that way.

For families, that means we need systems that aren't just smart, but also self-aware — tools that are checked, challenged, and changed when something's not working fairly.

It's explainable — in plain English

Let's be honest — one of the most confusing parts of AI is that it often feels like a mystery. It makes decisions, but we don't always know how or why. Inclusive AI doesn't hide behind jargon or complicated code. It explains itself clearly, especially when it affects people's lives.

What it looks like:

A school system uses AI to help place students in different learning groups. But instead of just giving you the result, it shows you exactly why your child was placed there, what kind of data was used (like past test results or classroom activity), and how to ask for a review if something doesn't seem right.

Why it matters:

Parents shouldn't need a tech degree to understand what's happening with their child's education or support. When AI decisions are clear and open, it helps families feel included — not sidelined. And when you know how a system works, you can speak up if something doesn't feel fair.

It keeps humans in the loop

No matter how smart AI gets, it's still just a tool — and like any tool, it can get things wrong. That's why inclusive systems are built with real people in mind. They don't force you to go along with whatever the computer says. Instead, they make sure there's always a way to ask questions, talk to someone, or just say, "This isn't working for me."

What it looks like:

Imagine you're using a government website to apply for support. An AI chatbot tries to help by sorting your request. But if it gets confused — or if your situation doesn't fit the standard form — you can easily click a button to speak to a real person. No dead ends, no endless loops, just help when you need it.

Why it matters:

AI should make life easier — not more frustrating. And when something as important as education, health, or support is on the line, families deserve to know they can press pause, ask for help, or take a different path. Keeping humans involved means keeping the system fair, flexible, and truly supportive.

It supports connection — not just efficiency

AI can be great at speeding things up — sorting, scanning, recommending. But life isn't just about being fast or efficient. It's about people, relationships, and care. The best AI systems understand this. They don't try to replace human connection — they support it

What it looks like:

Picture a mental health app your teenager uses to track their mood. The AI notices patterns — like a drop in sleep or more anxious days — and gently alerts a trusted adult, like a school counsellor or GP. It doesn't try to "diagnose" or make big decisions on its own. Instead, it helps the right people step in at the right time.

Why it matters:

Especially with kids and teens, trust and empathy are everything. A machine might be clever, but it can't sit with you, listen, or really understand what's going on in your world. Inclusive AI knows its role — and knows when to hand things over to a human who can truly help.

Inclusive AI doesn't mean perfect tech. It means thoughtful tech.

It means building tools that understand their limits — and are designed to keep learning, just like the people they serve. For families, that shows up in systems that are easier to understand, fairer in how they treat people, and better tuned to the real world — with all its differences, challenges, and strengths.

Because when AI is built to include — not just to perform — it becomes something families can trust. Not because it's perfect, but because it's built with care. It listens. It explains. And it remembers that behind every data point is a person.

What parents can look out for

Let's be honest: most of us aren't sitting around reading AI ethics papers in our spare time. We're packing lunches, helping with homework, juggling work, and trying to figure out why the school app just changed — again.

But even if you don't know all the technical ins and outs, you can spot whether an AI tool seems to be doing the right thing. You've probably already done this — sensed when something didn't feel fair, clear, or helpful. That gut feeling matters.

You don't need to become the family tech detective. But you can be curious. You can ask questions. And you can advocate for your child when something doesn't sit right.

Often, just naming the discomfort — "This doesn't feel clear" or "Can someone explain this?" — opens a conversation others were also waiting to have.

And remember: when schools, services, and companies know that families are paying attention, it encourages them to do better — to design tech that's not just smart, but also fair, respectful, and genuinely useful.

Because that's the kind of AI we want shaping our kids' futures.

How we can close the gap

It's easy to feel like the digital divide is just too big — too technical, too political, or too far removed from daily life. But here's the truth: closing that gap isn't only the job of governments, schools, or tech companies. It's something we can all be part of.

Because inclusion doesn't happen by accident. It happens when we pay attention. When we ask who's being left out — and make space for their voices. When we design systems that reflect real lives, not just ideal ones.

And while we might not control the whole system, we can make a difference in our own communities, schools, and conversations.

Here's how we help bridge the gap — and build a future where AI doesn't just work, but works for every family:

Talk about inclusion — not just access

These days, bridging the digital divide isn't just about whether families can get online — it's about what happens once they're there. Who does the technology work for? Who gets left behind? And who was never included in the first place?

We're at a point where we need to move the conversation beyond cables and connections. Inclusion means asking: is this tool fair, respectful, and usable for families from all walks of life — not just the tech-savvy or well-resourced?

What you can do:

Next time a new app, platform or AI tool is introduced at school or in services your family uses, don't be afraid to ask:

- Who was this designed for?
- Has it been tested with a wide range of families — including ours?
- Can everyone access it, understand it, and give feedback?

It's not about being critical. It's about making sure everyone's included from the start.

Share lived experience

Most tech decisions — whether in schools, services, or government — are made by people who may not fully understand the day-to-day reality of parenting. That's why your voice matters.

When families share their stories — the frustrations, the wins, the barriers they've hit — it helps decision-makers see the whole picture. What feels like a small example can highlight a much bigger issue that others are also experiencing.

What you can do:

If something's not working — or surprisingly helpful — speak up. You don't need a formal complaint or perfect words. A quick chat with your child's teacher, a message to your local MP, or a conversation at a community meeting can plant a seed for change.

Because lived experience isn't just personal — it's powerful.

Support community-based digital inclusion efforts

Some of the most important work to close the digital divide isn't happening in boardrooms or Parliament — it's happening right in our neighbourhoods.

Local libraries, Indigenous knowledge centres, migrant hubs, youth services, and community organisations are often the ones providing real help: internet access, device loans, digital skills training, and a safe place to ask questions without feeling judged.

What you can do:

If there's a program like this in your area, support it however you can — whether it's spreading the word, attending events, or simply recommending it to other families. If not, ask your council or school what's available. And when decisions about funding come up, speak up for these services — because they're doing the groundwork that helps families stay connected and confident.

Inclusion doesn't always start with policy. Sometimes, it starts with a neighbour, a shared computer, and a welcoming space.

Encourage diverse voices in tech

AI tools don't just appear out of nowhere — they're designed by people. And the backgrounds, experiences, and worldviews of those people shape what the technology pays attention to — and what it misses.

That's why diversity in tech isn't a bonus — it's essential. When we have more voices at the table — including women, First Nations peoples, regional Australians, neurodivergent thinkers, and those from different language or income backgrounds — we get technology that's more thoughtful, fair, and useful for everyone.

What you can do:

Look for programs that encourage a broader range of kids to see themselves in tech — not just coding, but also design, problem-solving, and ethical thinking. Support initiatives that open up STEM and digital careers to students who might not see it as "for them."

And when you're talking to schools, ask the bigger questions:

- Who's being encouraged to explore these paths?

- What role models are they seeing?
- Are we building a pipeline that reflects the real diversity of Australian families?

Because AI that works for everyone needs to be built by everyone — not just a narrow slice of society.

Ask for transparency and accountability

When technology plays a role in decisions that affect your family — like your child's learning, your health care, or your ability to access support — you deserve to know how it works. AI shouldn't be some mysterious black box. It should be clear, fair, and open to questions.

That's what we mean by transparency and accountability. It means the people using AI — whether in schools, hospitals, or government agencies — need to be upfront about what it's doing, how it was tested, and what happens when it doesn't get things right.

What you can do:

Don't be afraid to ask the important questions:

- Is this decision being made or influenced by AI?
- How do we know it's fair?
- Can I speak to a person if something seems off?
- Who's responsible if it makes a mistake?

You don't need to be an expert to ask these things. You just need to care — and as a parent or community member, your voice matters.

Asking questions doesn't mean you're being difficult. It means you're paying attention. And that helps make sure AI tools are held to high standards — not just for your family, but for every family they touch.

A shared future

Ultimately, inclusion in AI is about values. It's about building systems that reflect care, fairness, and respect for all families — not just the loudest voices or the most connected communities.

When we pay attention to the gaps — and take action to close them — we don't just improve the technology. We shape a digital future that's more just, more generous, and more ready for the kind of world we want our kids to grow up in.

And that's something worth showing up for. Not just as parents, but as a community.

Chapter Nine: Helping Kids Thrive — Tools, Tips & Conversations

Bringing it all together

By this point in the book, we've explored how AI shows up in schools, workplaces, homes, and even the games our kids play. We've looked at where it helps, where it complicates things, and where it might quietly shape behaviours in ways we don't always see.

So let's take a breath and bring it back to what really matters:

You don't need to be a tech whiz to raise kids who are smart about AI.

You don't have to memorise definitions or decode algorithms. What you do need—and already have—is the ability to guide, question, and connect. If you're reading this book, it's a sign you care enough to learn alongside your child — and that matters more than any technical know-how.

This chapter isn't about a list of things to worry about. It's about confidence. It's about turning "I'm not sure what this is" into "Let's figure it out together."

Here's what counts most: not your technical know-how, but your willingness to stay curious and engaged. Kids don't expect us to know everything—they just need us to be there, to ask thoughtful questions, and to show that it's okay not to have all the answers.

Start with Curiosity, Not Control

AI can feel overwhelming — partly because it's everywhere, and partly because it often runs quietly in the background. As a parent, your first instinct might be to block it, ban it, or try to stay one step ahead.

But here's another approach: get curious with your kids. Not just about the tech, but about how they're using it, why they like it, and what it might be doing under the surface.

You don't need a lecture, and neither do they. What you need is a conversation — one that invites them to notice what's happening behind the screen.

Try these openers:

"That's a cool game — do you think it learns how you play?"

"Why do you think YouTube keeps showing you those kinds of videos?"

"Do you think that quiz actually knows how smart you are — or is it guessing?"

These aren't trick questions. They're invitations. — When kids start noticing how systems work, they can begin thinking more critically. That's the real skill we're nurturing: not just using AI, but understanding it.

And if they ask something you don't know? That's okay. Say so. Say, "I don't know — let's look it up." That moment of learning side by side is more powerful than any perfect explanation.

Curiosity is your superpower as a parent. It builds trust, keeps the door open, and teaches kids that it's okay not to know — as long as you care enough to ask.

Age-by-age tips for talking about AI

Every child engages with technology a little differently — and what makes sense to a teenager won't always land with a five-year-old. That's why it helps to adjust how we talk about AI based on where kids are at developmentally.

You don't need a script. Just a few simple ways to make AI part of everyday conversation — without the eye rolls (or the blank stares).

Young kids (5–9): keep it simple and playful

At this age, kids are naturally curious — and they're often using AI without realising it. A talking toy, a voice assistant, or a YouTube recommendation — it's all part of their world.

What to say:

- "This toy listens to your voice so it knows what to say back."
- "That's a smart speaker — it follows instructions, kind of like a robot friend."
- "Have you noticed this game changes depending on how you play?"

What to encourage:

- Asking questions like "How does it know that?"
- Wondering out loud about what's real vs. what's programmed
- Learning through play — drawing what they think a robot does, or guessing how a character makes decisions

At this stage, you're planting the seeds for digital awareness — just like you might talk about where food comes from or how traffic lights work.

Tweens (10–13): Start exploring "how" and "why"

Tweens are becoming more independent online — using learning tools, gaming with friends, and watching content across platforms. It's the perfect time to introduce ideas like algorithms, personalisation, and even bias — in ways that feel relevant.

What to say:

- "YouTube recommends videos based on what you've watched before — but is that always a good thing?"
- "That game learns from your moves — do you think it makes it harder on purpose?"
- "Have you ever noticed ads showing up after you search something?"

What to encourage:

- Thinking about who creates the tools they use
- Asking "What's the goal of this app?" or "Is this fair for everyone?"
- Talking about privacy — what they share, and who might see it

This age is all about noticing patterns and building confidence to ask questions. They might not always answer, but they're definitely listening.

Teens (14+): Go deeper — ethics, bias, and real-world impact

Teenagers are often already using AI in sophisticated ways — sometimes more fluently than adults. But they may not see the bigger picture yet: how AI shapes opportunity, fairness, or identity.

What to say:

- "Do you trust the way that learning app scores you?"
- "How do you feel about companies using AI to scan CVs?"
- "What do you think AI should never be used for?"

What to encourage:

- Exploring AI in art, music, design — not just school
- Questioning how algorithms influence what they see, do, or buy
- Discussing issues like bias, misinformation, and automation — especially around jobs and future pathways

With teens, the goal isn't to lecture — it's to open space for their opinions. Often, they already have strong views. They just need an adult who's willing to listen (and maybe challenge them gently now and then).

In all age groups, the key is connection. Keep it conversational, not corrective. Be open about what you don't know. And remember — you're not just guiding them through tech. You're modelling how to think, question, and care in a changing world.

Does AI change the way we think?

One of the big worries about AI — especially in learning or decision-making tools — is that it can slowly chip away at kids' ability to think deeply, ask questions, and handle uncertainty.

And the tricky part? It doesn't always look like a problem. It can actually feel helpful:

- Search engines finish your sentence before you've even asked the question.
- Learning apps give instant answers, so there's no need to stop and think.
- Streaming services know your mood and suggest just the right thing.
- Voice assistants give quick replies — but often without much detail.

It's fast. It's easy. But when technology does too much of the thinking for us, kids miss out on chances to practise curiosity and problem-solving.

Over time, this can lead to:

- Relying on quick answers instead of exploring ideas
- Avoiding tricky questions or grey areas
- Struggling with problems that don't have clear steps
- Feeling less confident in their own thinking

These changes happen gradually — not all at once. But they matter.

That's why our job as parents isn't to stop kids from using AI. It's to help them stay connected to their own thoughts. To slow down when something feels too smooth. To wonder, "Why did it show me that?" or "Is that really true?"

Because the most powerful tool our kids have isn't an app — it's their own ability to think things through.

But there's another perspective — one we don't talk about as much:

What If AI Could Actually Boost Critical Thinking?

I've had moments where using AI helped me think more clearly — not because it gave me the answer, but because it helped me slow down and reflect. It pointed out things I hadn't considered, asked better questions, and helped me see where my thinking had gaps.

That's not replacing critical thinking — that's encouraging it.

And this is something we can offer our kids, too.

In a world full of headlines designed to stir emotion, social media that feeds us more of what we already believe, and endless content to scroll through, AI can actually be a way to pause and think. Not because it's always right — but because it helps us ask better questions.

In fact, research — and real student behaviour — is showing us this is already happening.

In a recent podcast conversation, educators looked at how university students are actually using AI. They studied over a million anonymous interactions with an AI tool. What they found was surprising:

Many students weren't just asking for answers. They were using AI to think more deeply:

- To explain tricky concepts
- To test ideas
- To get feedback on their work
- To ask, "What am I missing?"

One student used AI to debug a coding problem. Another used it to quiz themselves — then asked for harder versions. Others compared how different news sources framed the same story, to check for bias.

The most interesting part? Many were using AI to co-create — starting with a rough idea and working with the tool to shape it. Not replacing their thinking, but expanding it.

That's the kind of thinking we want to encourage:

- "Why am I being shown this?"
- "Did I choose this, or was it chosen for me?"
- "What's missing from this version of the story?"
- "Is this helping me understand — or just keeping me clicking?"

When kids learn to ask those kinds of questions — even just now and then — they're building more than digital skills. They're building awareness. They're learning how to think for themselves, even when technology is doing a lot in the background.

And the best part? You don't need a lesson plan. These habits often come from small, everyday moments — watching something together, trying out a new app, or just wondering aloud.

So yes, AI can absolutely be misused or overused. But it can also be a surprisingly helpful tool — not to do the thinking, but to support it.

And that's exactly what our kids need.

Tools and apps that support digital wisdom

Let's be honest — tech is everywhere, and not all of it is great for growing minds.

Some tools spark creativity, curiosity, and problem-solving. Others are designed to keep kids tapping and scrolling, without much thinking in between.

And as a parent, it's not always easy to tell which is which — especially when you don't have time to test everything yourself.

But there are some simple signs you can look for.

Instead of focusing on whether something is "educational" or "fun," try asking:

- Does this help my child make something, or just watch?
- Is there space for trial and error — or is it all about fast rewards?

- Can they change how it works to suit them, or is it one-size-fits-all?
- Does it explain its choices, or just deliver results with no context?

These small clues can tell you a lot about whether a tool is helping your child think, or just keeping them busy.

Because the best tech doesn't just hold attention — it supports reflection, confidence, and growth.

Tools that encourage creation over consumption

Kids light up when they get to make something — whether it's a silly cartoon, a short song, a mini game, or a story they've dreamed up. These kinds of tools do more than entertain. They give kids a sense of ownership, creativity, and confidence.

Look for tools that let your child:

- Build a story or comic from scratch
- Mix their own music or sounds
- Animate characters or code a simple game

Some great examples:

- Scratch — a free coding app where kids can make games, stories, and animations
- BandLab for Kids — a music mixing tool that's fun and intuitive
- Story-building apps or websites that invite kids to invent and narrate their own tales

Why it matters:

When kids are creating, they're actively using their imagination — and learning how to shape the technology they use. It shifts the mindset from "What can I watch?" to "What can I make?"

And that's a big step towards digital confidence.

Apps that explain themselves

Ever wondered why a certain video popped up, or how an app decided what your child should do next? Some tools now offer small explanations — little clues about how they work behind the scenes. It's not always detailed, but it's a helpful step.

What to look for:

- A "Why am I seeing this?" button next to videos, ads, or recommendations
- Settings you can adjust to influence what gets shown
- Learning apps that explain how they picked a level or gave a score

Why it matters:

When kids understand how an app works, they're less likely to take everything at face value. They start to realise: "Oh, this isn't magic — it's following a pattern." That kind of awareness helps them stay curious, ask better questions, and think more critically about the tech they use every day.

Games and apps that get kids thinking (not just tapping)

Not all screen time is mindless — but some apps are definitely more thoughtful than others.

The difference often comes down to how the tool is used. Is it helping kids solve problems, try different approaches, and think things through? Or is it just about collecting quick rewards and moving on?

What to look for:

- Games with puzzles that get trickier over time
- Activities that allow mistakes, retries, and different ways to solve a challenge
- Learning tools that encourage experimenting — not just choosing the "right" answer

Why it matters:

When kids have to pause, rethink, or try again, they're actually learning something valuable: persistence. They're building the confidence to tackle problems — even when the answer isn't obvious right away.

A bit of frustration isn't always a bad thing. In fact, it's often where the best learning happens.

Tools that let kids (and parents) take the wheel

Some apps just take over — they decide what your child sees, hears, or does, with little room for personal choice. Others give a bit more freedom — and that can make a big difference.

Look for tools that let you:

- Turn off autoplay or limit notifications
- Choose what kinds of content show up
- Adjust privacy settings or review activity
- Create separate profiles or settings for each child

Why it matters:

When kids (and parents) can customise how a tool works, it turns tech from something that runs them into something they run. It opens up chances to talk about how and why tools are set up a certain way — and how to tweak them to suit your family.

It's not about locking everything down — it's about making sure kids feel like they're in the driver's seat, not just along for the ride.

Tools that allow you to pause, rewind, or check

Some of the most helpful apps aren't the ones with flashy features — they're the ones that give kids space to slow down and reflect.

Especially with learning or AI-powered tools, it's great when there's a way to stop, rewind, or check how a decision was made.

What to look for:

- Clear explanations of why something was marked right or wrong
- Options to "undo," go back, or try again
- Features that let kids double-check their answers before submitting

Why it matters:

These little moments of pause can teach kids something powerful: you're allowed to think twice. When they can trace how a result happened — or fix a mistake — they're learning that good thinking takes time. And that even with smart tools, their judgment still matters.

Because the goal isn't just to get the answer — it's to understand how they got there.

So what can parents actually do?

You don't need to monitor every app or hover over your child's shoulder. But asking a few simple questions can make a big difference — not just for you, but for your child's own awareness.

Try asking:

- "What do you like about this app or game?"
- "Does it let you do things your way?"
- "Can you tell when it's using AI?"
- "Do you feel like you're in control — or is it kind of deciding for you?"
- "Do you ever find it hard to stop using it — even when you want to?"
- "Do you think it's designed to keep you playing or scrolling?"

These aren't trick questions. They're ways to start a conversation and help your child reflect on how they use technology — and how it might be using them.

Because in the end, it's not about the flashiest features or the smartest software. What really matters is whether the tool leaves room for:

- Thinking (not just reacting)
- Creating (not just consuming)
- Connecting (not just scrolling)

The best tech helps kids grow — not just keep them busy.

Everyday conversations that build confidence

In earlier chapters, we looked at how to ask good questions, notice patterns, and stay curious. But if there's one thing that really helps kids develop confidence around AI, it's not more tech skills.

It's knowing they can come to you — even when they feel unsure, awkward, or wrong.

Because the truth is, AI can make things feel a bit fuzzy. Answers come quickly. Choices are suggested before we've had time to think. And sometimes, kids don't know what to make of what they've just seen, heard, or clicked.

But how do we build something steady and simple: a sense of safety.

Confidence grows through connection

In earlier chapters, we talked about the power of good questions, noticing patterns, and staying curious. But when it comes to building real confidence — especially around AI — there's something even more important than tech skills:

It's knowing they can come to you.

Even when they feel unsure.

Even when they clicked on something weird.

Even when they got something wrong.

Because let's be honest: AI tools often feel fast and fuzzy. Answers appear before we've finished the question. Suggestions pop up before we've had a chance to think. And sometimes, kids just feel confused or caught off guard — but don't quite know how to talk about it.

So how do we help them feel safe in all of that?

By offering something simple and steady: a space where they know they can ask, wonder, and make mistakes without being shut down or judged.

That's what builds trust. That's what builds confidence. And it often starts with the smallest conversations — the ones that say, "I'm here, and we'll figure it out together."

You don't have to be the expert

You're not expected to have all the answers — seriously.

Sometimes the best thing you can say is, "I don't know either." Not as a cop-out, but as an honest starting point.

Because when your kid sees you stop and think something through, or admit you're not sure, it shows them that it's normal to pause and figure things out. No need to bluff or guess or pretend you're in control of everything.

Try something like:

- "Let's check it out together — I'm curious now too."
- "That's a good point. What do you reckon it means?"

These aren't big teaching moments. They're just everyday conversations — but they help build the kind of thinking that lasts.

Reflection, not perfection

AI tools can make everything feel fast — like there's no need to stop and think. But those small pauses? They matter.

A quick question like:

- "Would you do anything differently next time?"
- "Did that feel like your choice, or was the app kind of steering?"

That's often enough to spark a bit of deeper thinking. Not to catch them out — just to remind them that it's okay to change their mind or ask another question.

Sometimes, the most helpful thing we can teach isn't what to think — it's that they don't have to rush.

Keep the door open — and relaxed

You've probably noticed by now: the real magic isn't in the perfect tool or perfect answer. It's in the everyday habit of staying curious — and staying connected.

That's what helps kids feel confident asking questions, especially when things online don't quite sit right.

It doesn't have to be a big moment. Maybe it's your child pausing before clicking something, or telling you about a strange recommendation they got. What matters is that they feel they can bring it up — without judgement or panic.

Even a simple response like:

- "Yeah, that's worth thinking about."
- "Want to unpack it together?"

— keeps the door open. Not just for that chat, but for the next one too.

The heart of it all: raising humans, not just users

So we come back to the questions that is at the heart of this book - What kind of future are we preparing our kids for — and how can we help them thrive in a world shaped by AI?

There's no single roadmap. No perfect formula. And certainly no app that can do it for us. But if this book has offered anything, I hope it's this:

A reminder that you're not alone.

That you don't have to be a tech expert.

And that your presence — your questions, your care, your curiosity — already count for more than you might realise.

Because raising thoughtful, confident, values-driven humans has never been about keeping up with every new tool or trend. It's about:

- Staying present, even when things are messy or unclear
- Staying open, even when you don't have the answers
- And staying curious, especially when the world wants quick responses

We're not trying to raise kids who can just use AI.

We're trying to raise kids who can think alongside it, question it, challenge it — and ultimately lead in the world it's shaping.

That's not about knowing everything. It's about showing up — again and again — with the kind of thinking, empathy, and grounding that no machine can replicate.

Because that's what truly prepares them for the future.

Not the flashiest tech.

Not the right answers on the first try.

But the ability to slow down, ask better questions, and choose with intention.

In a world that's changing fast, the real goal isn't just raising kids who can keep up with technology — it's raising kids who strengthen everything that makes us human.

Empathy. Integrity. Imagination.

The things no machine can replicate.

And the things our future needs most.

A Note on AI and This Book

Since this is a book about AI, you might be wondering: did I use AI to help write it? The answer is yes — and I think it's important to say so clearly.

AI was especially useful in the research stage. It helped me scan through the huge amount of daily news, reports, and policy updates about AI, and quickly point me toward sources I might otherwise have missed. Instead of trawling through endless links, I could spend my energy on analysing, fact-checking, and deciding what really mattered for parents and families.

Every source you see referenced here has been checked manually, every example weighed, and every story written in my own voice. The AI never replaced my judgement — it simply sped up the process of finding the raw material. Think of it as a research assistant who can pull together a pile of articles in seconds, leaving me free to do the human part: questioning, interpreting, and shaping it into something meaningful.

I chose to use AI in this way because the best way to understand these tools is to work with them, reflect on the experience, and stay in control of the process. My hope is that this book shows how AI can support human work without taking it over — and why transparency about that matters.

So yes, AI played a role in the research and drafting process. But the perspective, the choices, and the responsibility for what you've read? That's all mine.

AI Terms & Tools:
A-Z for Curious Parents

A handy, plain-English guide to the AI-related terms, tools, and companies you're likely to hear — whether in headlines, school newsletters, or your child's latest "homework helper" app.

A

Agentic AI

AI that doesn't just respond to instructions — it can take actions, set goals, or make decisions on its own, often across multiple steps. Think of it as an AI that's more proactive, not just reactive.

AI Detector

A tool that tries to tell whether a piece of writing or content was created by a person or by an AI system. These detectors aren't very reliable — they can wrongly flag real student work as "AI-generated" or miss writing that actually used AI. Many schools and experts now suggest focusing on responsible use of AI instead of relying on these tools.

AI Literacy

Understanding how AI works, where it shows up, and how to use it wisely. It's like digital literacy, but focused on AI tools and their effects.

Algorithm

A step-by-step set of instructions a computer uses to solve a problem or make decisions — like deciding what video to recommend next.

Anthropic

A company behind AI tools like Claude — focused on safety and human alignment in AI.

Artificial Intelligence (AI)

Technology that mimics human thinking — learning, deciding, recognising patterns — but doesn't actually "think" like a human.

Artificial General Intelligence (AGI)

A future version of AI that could learn and reason like a human across many different areas — not just one specific task. It's often seen as the "next level" of AI, beyond what we have now.

Automation

When tasks are done by machines or software instead of people — like chatbots answering customer questions or software grading tests.

B

Bias (in AI)

When AI makes unfair or skewed decisions, often due to patterns in the data it was trained on. Important to watch for, especially in education, hiring, or law enforcement.

Big Data

Extremely large datasets that AI systems analyse to detect patterns or make predictions.

C

Chatbot

A computer program designed to have a conversation with a person. Some are basic (e.g., answering FAQs), while others are highly advanced.

ChatGPT

A popular AI tool created by OpenAI that can write, explain, or answer questions. It's like a very smart search engine — but it can also make mistakes.

Claude

An AI chatbot from Anthropic, similar to ChatGPT but with a focus on being helpful, honest, and harmless.

Computer Vision

AI that can "see" — it analyses images or video, like detecting faces in photos or objects in traffic cameras.

Content Moderation (with AI)

Using AI to scan and filter online content — like removing harmful videos or blocking offensive comments.

Copilot

A general term for an AI assistant that helps you do tasks — like writing, editing, coding, or creating presentations. Microsoft's Copilot is one of the most well-known, built into tools like Word and Excel.

Copilot (Microsoft Copilot)

Microsoft's AI assistant built into tools like Word and Excel. Helps write, summarise, or generate content.

D

DALL·E

An AI tool from OpenAI that turns text into images — just like Midjourney. You type a sentence, and it draws a picture to match. The name is a mashup of "Dali" (the artist) and "WALL·E" (the robot).

Data Privacy

The rights and practices around how personal information is collected, used, stored, and shared by digital tools and AI

Deepfake

Fake video or audio created using AI to look or sound like real people. Can be entertaining — or harmful if used to mislead.

Digital Footprint

The trail of data you leave behind when using online services — including how AI might learn from your clicks, views, or searches.

Digital Twin

A virtual version of something real — like a person, place, or system — used to model or test things. Not common in homes, but sometimes used in education or health.

E

Edge Computing

Technology where computing happens closer to the data source (like on a device) rather than a distant server; relevant when mobile or remote tools use AI offline.

Ethical AI

The idea that AI should be designed and used responsibly — with fairness, safety, and transparency in mind.

Explainable AI (XAI)

AI systems built to make their reasoning clear and understandable, so users can see why a decision was made.

F

Facial Recognition

AI that identifies people based on their face. Used in phones, surveillance, and some school safety systems — but also raises privacy concerns.

Fine-tuning

The process of adapting an AI model to work better in specific situations — like helping a learning tool better support reading or maths.

G

Generative AI

AI that creates content — from text to images to music. These tools can write stories, generate art, or even mimic voices.

Google DeepMind

A research company owned by Google that develops advanced AI models, including tools used in science, healthcare, and games.

GPT (Generative Pre-trained Transformer)

A kind of AI that's really good at reading and writing like a person. It powers tools like ChatGPT.

"Pre-trained" means it learned by reading lots of text before it ever talks to you.

"Transformer" is just the tech that helps it understand what you're asking and respond in full sentences.

Grok

An AI chatbot developed by xAI (Elon Musk's AI company) and integrated with X (formerly Twitter). It's designed to answer questions and join conversations — sometimes with a sarcastic or edgy tone.

H

Hallucination (in AI)

When an AI tool gives an answer that sounds right — but isn't. A common issue with tools like ChatGPT.

Human-in-the-loop

When people stay involved in decisions made by AI — like reviewing, approving, or stepping in when it really matters. It helps keep human judgement in the process.

I

Inference

The process of AI using what it knows to make predictions or decisions — like guessing what ad to show next.

L

Large Language Model (LLM)

A type of AI trained on huge amounts of text — it powers tools like ChatGPT and Claude. These models can answer questions, write stories, and more.

M

Machine Learning

A branch of AI where systems improve over time by learning from data — like recognising handwriting or predicting shopping habits.

Meta

The company behind Facebook, Instagram, and some large AI models like LLaMA.

Midjourney

An AI tool that creates images based on written prompts. You type in a description ("a koala in space," for example), and it generates a picture to match. Often used by artists, designers, and curious kids.

Model Drift

When an AI tool starts making worse guesses over time because the world has changed — like if it was trained on old data that no longer fits how people live or learn today.

N

Natural Language Processing (NLP)

How AI understands and responds to human language — used in chatbots, translation apps, and virtual assistants.

O

On Device AI

AI that works directly on your phone, tablet, or laptop — even without internet. Useful for privacy, speed, or when access to the cloud is limited.

OpenAI

The company behind ChatGPT. Its goal is to create AI that's useful and safe for everyone.

P

Personalisation

When AI tailors what you see — like playlists, ads, or search results — based on your past behaviour.

Perplexity

An AI-powered search tool that gives answers in full sentences instead of just links. It works a bit like ChatGPT, but focuses on finding and summarising information from the web — and usually shows its sources.

Prompt

The message or question you type into an AI tool to get a response — like "Write a bedtime story about koalas."

Prompt Engineering

The skill of writing clear, specific inputs (prompts) to get better responses from AI tools.

R

Recommendation Engine

The AI behind "you might also like" suggestions — used in YouTube, Netflix, shopping apps, and more.

Reinforcement Learning

A way AI learns by trial and error — it's how some AI systems learn to play games or drive virtual cars.

Responsible AI

A general term for designing and using AI in ways that are fair, safe, and aligned with human values.

S

Sentiment Analysis

When AI tries to figure out how someone feels — happy, sad, frustrated — based on what they write or say. Used in customer service and education apps.

Singularity

A theoretical future point where AI becomes smarter than humans — not something we're close to, despite media hype.

Stability AI

Makers of Stable Diffusion, a popular image generation model — widely used for AI art.

Superintelligence (or Artificial Superintelligence, ASI)

A theoretical form of AI that would surpass human intelligence in every way — able to think, learn, and solve problems far beyond our abilities. It doesn't exist today, but experts debate what it could mean for the future if it ever does.

Supervised Learning

A way AI is trained using labelled examples — like showing it thousands of photos of cats and dogs to help it tell them apart.

Synthetic Media

Any image, video, or sound created by AI — including avatars, fake voices, and deepfakes.

T

Token

A token is a small piece of text — like a word or part of a word — that AI tools use to understand and build sentences.

The AI tools, like ChatGPT, don't see a complete sentence at once. They look at one token at a time and guess what should come next — like putting a sentence together, piece by piece.

Training Data

The information AI learns from — the books, articles, videos, or images it studies before being used.

Transparency (in AI)

How open and understandable an AI system is — important for knowing what decisions it's making and why.

V

Voice Assistant

AI tools that respond to voice commands — like Siri, Alexa, or Google Assistant.

Z

Zero-shot Learning

When an AI tool can do something new with little or no examples — like answering a question it hasn't seen before. This helps tools adapt to new situations quickly, but also means they can get things wrong.

Further Reading & Resources

For parents who want to keep learning — or just feel a little more confident alongside their kids — here are some trusted, easy-to-understand resources.

This isn't a list of technical deep-dives. It's a collection of thoughtful, balanced tools I found helpful while researching and writing this book — and in my own learning as a parent and professional.

There's no shortage of AI content out there, but these stood out as genuinely useful starting points.

Australian Resources

Artificial Intelligence – Australian Government Overview
Central hub for Australia's AI strategy, initiatives, funding, and responsible tech development.

https://www.industry.gov.au/science-technology-and-innovation/technology/artificial-intelligence

Voluntary AI Safety Standard – 10 Guardrails
Outlines practical, people-first guidelines to help organisations design and use AI safely.

https://www.industry.gov.au/publications/voluntary-ai-safety-standard/10-guardrails

Australia's Artificial Intelligence Ethics Principles
A set of 8 core values — like fairness, privacy, and accountability — to guide ethical AI use.

https://www.industry.gov.au/publications/australias-artificial-intelligence-ethics-principles

National AI Capability Plan (Department of Industry)

Australia's plan to grow national skills, research, and responsible AI use across sectors.

https://www.industry.gov.au/news/developing-national-ai-capability-plan

Ministerial Media Release: AI Plan to Boost Capability

Official announcement outlining goals for Australia's national AI roadmap.

https://www.minister.industry.gov.au/ministers/husic/media-releases/australian-first-ai-plan-boost-capability

Australia's Artificial Intelligence Action Plan

The government's broader AI vision for responsible innovation and global leadership.

https://www.industry.gov.au/publications/australias-artificial-intelligence-action-plan

National AI Centre (CSIRO)

https://www.csiro.au/naic

Leads responsible AI development in Australia. Great for understanding national strategies, tools like the AI Impact Navigator, and real-world case studies.

Responsible AI Index 2024 (Fifth Quadrant)

Tracks how Australian businesses and sectors are progressing on ethical and inclusive AI practices.

https://www.fifthquadrant.com.au/responsible-ai-index-2024

eSafety Commissioner

https://www.esafety.gov.au

Excellent guides on online safety for children, teens, and families — including emerging AI risks and digital wellbeing.

Raising Children Network

https://raisingchildren.net.au

Trusted parenting advice backed by experts. Includes tech use, screen time, digital learning, and more.

Australian Human Rights Commission – AI and Rights

https://humanrights.gov.au

For those curious about the ethical side of AI — especially fairness, transparency, and discrimination.

Australian Framework for Generative AI in Schools

A national guide endorsed by all states and territories, offering principles for safe, ethical, and effective use of AI in schools.

https://www.education.gov.au/schooling/resources/australian-framework-generative-artificial-intelligence-ai-schools

CSER AI in Schools (University of Adelaide)

Provides free professional learning, lesson plans, and support for AI in classrooms.

https://csermoocs.adelaide.edu.au/professional-learning/artificial-intelligence-in-schools

New South Wales – Helping Parents and Carers Understand AI in Education

https://education.nsw.gov.au/teaching-and-learning/education-for-a-changing-world/helping-parents-and-carers-understand-ai-in-education-

New South Wales – AI in Education

https://education.nsw.gov.au/teaching-and-learning/education-for-a-changing-world/artificial-intelligence-in-education

Victoria – AI in Schools (Academy)

https://www.academy.vic.gov.au/resources/collections/ai-in-schools

Queensland – Generative AI in Queensland State Schools

https://education.qld.gov.au/initiativesstrategies/equityexcellence/digital-innovation-in-teaching-learning/Pages/generative-ai-in-schools.aspx

South Australia – AI in Schools: Information for Parents and Carers

https://www.education.sa.gov.au/parents-and-families/curriculum-and-learning/ai/artificial-intelligence-ai-in-schools-information-for-parents-and-carers

Global and Accessible Sources

Common Sense Media – AI & Families

https://www.commonsensemedia.org

US-based but globally useful. Offers clear reviews, explainer articles, and family-friendly guides on apps, AI, and digital tools.

UNESCO – AI in Education

https://www.unesco.org

A global look at how AI is being used in schools, and what it means for equity, access, and child rights.

BBC Bitesize & Tech Explained

https://www.bbc.co.uk/bitesize

Simple, school-level explanations of digital terms and trends – helpful for parents and older kids alike.

Podcasts

These podcast episodes and series offer accessible, thought-provoking conversations about AI, ethics, and the future – often with a human, parenting, or education lens.

They're not too techy, and many feature real stories and expert insights.

You can usually find them by searching the title on your preferred podcast app (Spotify, Apple Podcasts, Castbox, etc.).

AI for Kids Podcast

Hosted by Dipti, CEO of LittleLit, this podcast delves into AI education for kids, offering insights for parents and teachers on AI literacy and tools.

AI Literacy for Parents

A resource aimed at empowering parents to navigate the evolving world of artificial intelligence and its implications for their children.

Create, Learn & Tomorrow: The AI Education Show

This podcast bridges the gap between today's classrooms and tomorrow's AI-driven opportunities, offering insights for parents and educators.

AI Parenting Podcast

Dr. Ed Tse helps families transition from screen time to quality time in the AI era. The podcast offers practical advice on navigating AI's impact on family life.

AI Education Podcast

Hosted by Dan Bowen (Microsoft Australia) and Ray Fleming (InnovateGPT), this weekly show explores how AI is shaping education in Australia. It's approachable and informative for parents and educators alike.

Pioneers of AI

Hosted by AI scientist and author Rana el Kaliouby, this podcast features conversations with leading creators, critics, and thinkers about how artificial intelligence is changing the way we live.

Books

If you're the kind of parent who likes to dig deeper — or just enjoys learning through stories and thoughtful reflections — these books are a great place to start.

They cover AI, technology, and the future of work from different angles: some are personal, some practical, some eye-opening.

Man-Made by Tracey Spicer (2023)

Investigates how bias from the past is being built into the technologies shaping our future.

https://www.simonandschuster.com.au/books/Man-Made/Tracey-Spicer/9781761106378

Girl Decoded by Rana el Kaliouby (2020)

A memoir exploring the integration of emotional intelligence into technology through the development of Emotion AI.

https://www.penguin.com.au/books/girl-decoded-9780241451533

Unmasking AI by Joy Buolamwini (2023)

A powerful narrative on racial and gender bias in AI, from the founder of the Algorithmic Justice League.

https://www.penguinrandomhouse.com/books/719693/unmasking-ai-by-joy-buolamwini

The Worlds I See by Fei-Fei Li (2023)

A memoir from a leading AI researcher, exploring the human side of innovation, ethics, and scientific discovery.

https://us.macmillan.com/books/9781250878544/the-worlds-i-see

Brave New Words by Salman Khan (2024)

Explores how AI can revolutionize education and personalize learning experiences, offering insights for parents, educators, and policymakers.

https://www.penguinrandomhouse.com/books/740806/brave-new-words-by-salman-khan/

Made in the USA
Monee, IL
04 November 2026

67383949R00118